I0057686

Applied Thermal Measurements at the Nanoscale

A Beginner's Guide to Electrothermal Methods

Lessons from Nanoscience: A Lecture Note Series

ISSN: 2301-3354

Series Editors: Mark Lundstrom and Supriyo Datta
(Purdue University, USA)

"Lessons from Nanoscience" aims to present new viewpoints that help understand, integrate, and apply recent developments in nanoscience while also using them to re-think old and familiar subjects. Some of these viewpoints may not yet be in final form, but we hope this series will provide a forum for them to evolve and develop into the textbooks of tomorrow that train and guide our students and young researchers as they turn nanoscience into nanotechnology. To help communicate across disciplines, the series aims to be accessible to anyone with a bachelor's degree in science or engineering.

More information on the series as well as additional resources for each volume can be found at: http://nanohub.org/topics/LessonsfromNanoscience

Published:

Lessons from Nanoscience:
A Lecture Note Series

Vol. 7

Applied Thermal Measurements at the Nanoscale

A Beginner's Guide to Electrothermal Methods

Zhen Chen
Southeast University, China

Chris Dames
UC Berkeley

World Scientific

NEW JERSEY · LONDON · SINGAPORE · BEIJING · SHANGHAI · HONG KONG · TAIPEI · CHENNAI · TOKYO

Published by

World Scientific Publishing Co. Pte. Ltd.

5 Toh Tuck Link, Singapore 596224

USA office: 27 Warren Street, Suite 401-402, Hackensack, NJ 07601

UK office: 57 Shelton Street, Covent Garden, London WC2H 9HE

Library of Congress Cataloging-in-Publication Data

Names: Dames, Christopher Eric, author. | Chen, Zhen
  (Professor of mechanical engineering), author.
Title: Applied thermal measurements at the nanoscale : a beginner's guide to electrothermal methods /
Chris Dames (University of California at Berkeley), Zhen Chen (Southeast University, China).
Other titles: Lessons from nanoscience ; v. 7.
Description: Singapore ; Hackensack, NJ : World Scientific Publishing Co. Pte. Ltd., [2018] |
  Series: Lessons from nanoscience: a lecture notes series, ISSN 2301-3354 ; vol. 7 |
  Includes bibliographical references and index.
Identifiers: LCCN 2018021230| ISBN 9789813271104 (hardcover ; alk. paper) |
  ISBN 9813271108 (hardcover ; alk. paper)
Subjects: LCSH: Nanostructured materials--Thermal properties. | Temperature measurements. |
  Thermal conductivity.
Classification: LCC QC176.8.T4 D36 2018 | DDC 536/.2012--dc23
LC record available at https://lccn.loc.gov/2018021230

British Library Cataloguing-in-Publication Data
A catalogue record for this book is available from the British Library.

First published 2018 (Hardcover)
Reprinted 2019 (in paperback edition)
ISBN 978-981-121-298-7 (pbk)

Copyright © 2018 by World Scientific Publishing Co. Pte. Ltd.

All rights reserved. This book, or parts thereof, may not be reproduced in any form or by any means, electronic or mechanical, including photocopying, recording or any information storage and retrieval system now known or to be invented, without written permission from the publisher.

For photocopying of material in this volume, please pay a copying fee through the Copyright Clearance Center, Inc., 222 Rosewood Drive, Danvers, MA 01923, USA. In this case permission to photocopy is not required from the publisher.

For any available supplementary material, please visit
https://www.worldscientific.com/worldscibooks/10.1142/11009#t=suppl

Typeset by Stallion Press
Email: enquiries@stallionpress.com

To our families

Preface

Brief description

This book offers guidance for beginners to design and conduct measurements of thermal properties at the nanoscale using electrothermal techniques. The emphasis is on measuring thermal conductivity, though thermal diffusivity and thermal boundary resistance (also known as thermal contact resistance) are also touched on. The only prerequisites expected of the reader are the basic familiarity with electrical instruments typical of a bachelor's degree in science or engineering, and knowledge of the three basic heat transfer laws, namely, Fourier's law of heat conduction, Newton's cooling law of convection, and the Stefan–Boltzmann radiation law.

We have two goals in this book. The basic one is to introduce readers to some of the popular techniques for electrothermal measurements at the nanoscale, and guide them to conduct successful measurements by themselves. As a more advanced goal, we also hope to stimulate some readers to extend existing techniques, or even develop new methods from scratch, according to their own unique interests and constraints, e.g., sample geometry and thermal properties.

Organization of the book

Chapter 1 briefly introduces prevailing experimental techniques, and emphasizes the tension between simplicity of microfabrication and simplicity of heat transfer model.

Chapter 2 discusses how to select the proper technique, considering the sample's dimensionality and geometry.

Chapter 3 presents some detailed aspects of experimental design and validation which are usually omitted from journal papers, and even theses, but which are nevertheless important for preparing a successful experiment.

Chapter 4 highlights the importance of uncertainty and sensitivity analysis, and discusses both the conventional partial derivative method and a less familiar Monte Carlo scheme.

The appendices summarize additional notes, including a list of notation, the key settings of a lock-in amplifier, the effect of natural convection on the 3ω method, the advantages of a four-probe AC measurement, an op-amp circuit to convert a voltage source to a current source, the vacuum level needed to suppress air conduction, radiation shields to minimize radiation losses, and brief comments on material properties and the lognormal distribution.

Distinction from related works

There are already many review articles and book chapters digging deeply into the technical details of measurement techniques for thermal properties of nanostructures. Early reviews on thin films include Refs. [1–3]. A recent extensive collection of reviews, edited by Chen [4], refreshes perspectives on a broad range of experimental techniques for various nanostructures, including electrothermal methods for measuring thin films [5] and nanowires and nanotubes [6].

This book takes a higher level, more pedagogical approach, and is aimed at a relatively less expert audience than a typical review article. Our hope is that a motivated early-stage graduate student could use this book to select the best measurement technique for their sample, and in consultation with more specialized literature, develop their own experimental setup and see through a successful measurement with confidence.

References

[1] D. G. Cahill, H. E. Fischer, T. Klitsner, E. T. Swartz and R. O. Pohl, "Thermal conductivity of thin films: Measurements and understanding," *J. Vaccum Sci. Technol.* **7**, p. 1259, 1989.

[2] K. E. Goodson and M. I. Flik, "Solid layer thermal-conductivity measurement techniques," *Appl. Mech. Rev.* **47**, pp. 101–112, 1994.

[3] T. Borca-Tasciuc and G. Chen, "Experimental techniques for thin-film thermal conductivity," in *Thermal Conductivity: Theory, Properties, and Applications*. Springer, pp. 205–237, 2004.

[4] G. Chen, "Probing nanoscale heat transfer phenomena," *Annu. Rev. Heat Transf.* **16**, pp. 1–6, 2013.

[5] C. Dames, "Measuring the thermal conductivity of thin films: 3 omega and related electrothermal methods," *Annu. Rev. Heat Transf.* **16**, pp. 7–49, 2013.

[6] A. Weathers and L. Shi, "Thermal transport measurement techniques for nanowires and nanotubes," *Annu. Rev. Heat Transf.* **16**(1), pp. 101–134, 2013.

Acknowledgments

We would like to thank Tim Fisher for introducing us to the *Lessons from Nanoscience* series. Our journey from that introduction to this finished product has been a long one, and we appreciate the patience of Mark Lundstrom and the editorial staff.

We would also like to acknowledge many of our colleagues for fruitful discussions. In particular, Z.C. acknowledges his colleague at Southeast University, Juekuan Yang, for pointing him towards several useful references on the suspended microfabricated-device method. Z.C. also acknowledges his former labmate at UC Berkeley, Sean Lubner, for sharing his notes on the voltage to current conversion circuit. Z.C. finally acknowledges his current student, Nan Chen, for carefully proofreading the manuscript. C.D. is grateful for the many wonderful students, postdoctoral scholars, and collaborators he has worked with over the years, whose ideas and insights influence this book throughout. In particular he would like to thank Wanyoung Jang, whose skill and tireless efforts at microfabrication and experimentation were essential to several key systems which pervade this book.

We finally would like to thank our families. Most of the time spent on this book was stolen from them. Without them, we could have finished this book much earlier; but without them, it would be meaningless for us to work hard on anything.

Zhen Chen & Chris Dames

Contents

Nomenclature

$\mathbf{a}$	Physics variables (usually a vector including many components)
a	One component of $\mathbf{a}$
A	Area
b	Heater half-width (with subscript "sens": Sensor half-width)
c	Uncertainty contribution
C	Volumetric heat capacity
d	Distance
D	Thermal diffusivity
h	Heat transfer coefficient for conduction, convection, or radiation
I	Current
k	Thermal conductivity
l	Heater length
L	Sample length (with subscript p: Penetration depth)
m	Slope
n	Length of $\mathbf{Z}$, or number of data points (clear from context)
N	Number of experiments or MC repeats (clear from context)
p	Pressure
P	Perimeter
Q	Heat flow
R	Thermal resistance (with subscript e: Electrical resistance)

S Sensitivity

t Thickness or time (clear from context)

T Temperature

u Uncertainty

w Width of sample

V Voltage

$\mathbf{X}$ Control variables (usually a vector including many components)

$\mathbf{Y}$ Response variables (usually a vector including many components)

$\mathbf{Z}$ Concatenation of $\mathbf{X}$ and $\mathbf{Y}$, i.e., $\mathbf{Z} = [\mathbf{X}; \mathbf{Y}]$

z One component of $\mathbf{Z}$

Z Thermal impedance (with subscript e: Electrical impedance)

Greek

α Temperature coefficient of electrical resistance

β Fin parameter (β^{-1} is the characteristic decay length of a fin's $T(x)$ profile)

γ Dimensionless ratio in the T-bridge method, $\gamma = R_{\text{Htr}}/(4R_{\text{sampl}})$

δ Perturbation of variables (Chapter 4)

ε Emissivity

η A constant (≈ 0.923)

ξ Square root of the ratio between the in-plane and cross-plane thermal conductance

ρ Electrical resistivity

σ Electrical conductivity or Stefan–Boltzmann constant (5.67×10^{-8} $\text{Wm}^{-2}\text{K}^{-4}$) (clear from context)

τ Time constant (with subscript e: Electrical time constant)

ϕ Phase angle

χ Used in definition of confidence interval: $\text{CI} = (1 - \chi) \times 100\%$

ω Angular frequency (with subscript H: heating frequency, $\omega_H = 2\omega$)

Subscripts and Superscripts

$''$	Area normalized
$'''$	Volume normalized
∞	Environment
0	Condition of negligible self heating (e.g., $R_{e,0}$, when driving current $\to$ 0)
$1\omega,\ 3\omega$	Harmonic number
I, II	Two principal axes
$\perp$	Cross-plane direction
$\parallel$	In-plane direction
avg	Average
bkg	Background
B–G	Bloch–Grüneisen
BotOx	Bottom oxide layer
c	Contact
CC	Chip carrier
CF	Cold finger
cr	Cross sectional, e.g., A_{cr} is cross-sectional area
char	Characteristic
cond'n	Conduction
conv'n	Convection
cylind	Cylindrical tube
e	Electrical
eff	Effective
expt	Experiment
film	Thin film
gr	Graphene flake
gr–ox	from graphene flake to top/bottom oxide layer
gr–Si	from graphene flake through the bottom oxide layer to Si substrate
H	Heating
High-T	High temperature limit
Htr	Heater
HtrToS1	From heater to sensor #1
HtrToS2	From heater to sensor #2

i	ith component
island	Heating or sensing island in the microfabricated-device method
LB	Lower bound
linear	Linear fit
max	Maximum
min	Minimum
msrd	Measured
ox	Oxide layer
ox–Si	From bottom oxide layer to Si substrate
OxUnderHtr	The portion of the top oxide layer beneath the line heater
p	Penetration
rad'n	Radiation
rms	Root-mean-square
s1	Sensor #1
s2	Sensor #2
sampl	Sample
Sens	Sensor
Serpentine	Serpentine heater in the microfabricated-device method
set	Set point
Si	Silicon substrate
Si-CF	From Si substrate to cold finger
silver	Silver paint
sim	Simulated
SLG	Single layer graphene
sub	Substrate
surf	Surface
synth	Synthesized
TopOx	Top oxide layer
true	True physics
UB	Upper bound
x	in-phase component of the current or voltage
y	out-of-phase component of the current or voltage

List of Figures

List of Tables

Chapter 1

What is in Your Toolbox?

"Intuition is nothing more than having studied all the important solved problems."

— S. D. Senturia
(inspired by G. Polya's How to Solve It)

1.1. Introduction

Nanoscale heat transfer research has been largely driven by applications in information technology and energy. For example, as transistors and memory relentlessly shrink in size and accelerate in speed, better solutions for heat dissipation are required. Good thermal design relies on accurate knowledge of micro/nanoscale devices' thermal properties. However, the thermal properties of nanostructures can differ dramatically from their bulk counterparts. For example, the thermal conductivity (k) of a silicon nanowire or a nanocrystalline silicon substrate may be over an order of magnitude lower than k of an intrinsic silicon wafer, depending on the diameter of the nanowire or the grain size of the nanocrystal [1, 2]. Thus accurately measuring the thermal properties of such nanostructures is mandatory.

Thermal property measurements, whether of lumped thermal conductance, thermal conductivity, or thermal contact resistance, are fundamentally built on measurements of temperature and heat flux. Among the myriad ways to heat a sample and to measure its temperature, this book focuses exclusively on electrical techniques: Joule heating and resistance thermometry. Such electrothermal methods are convenient and relatively easy to try because of the

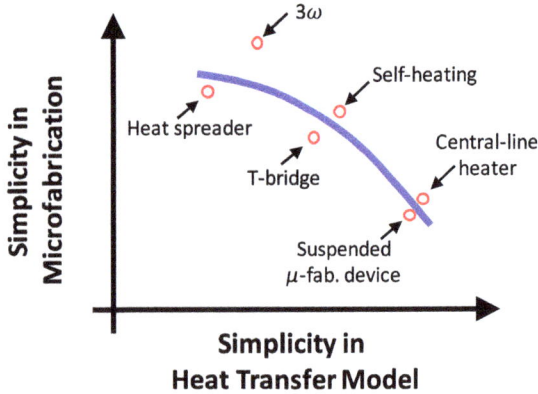

Fig. 1.1   The conceptual tradeoff (blue solid line) between simplicity of micro-fabrication and heat transfer model. Some popular techniques (red symbols) are depicted qualitatively, and will be discussed in detail in this book. See also Table 1.1.

widespread availability of high-accuracy instruments at relatively low cost, and the ease of automating instrument control and data acquisition directly in the electrical domain. However, for nanoscale samples the attendant microfabrication challenges are not to be overlooked, and indeed the tradeoff between simplicity of fabrication and simplicity of thermal analysis is one of the major themes of this book.

An ideal measurement technique of course is both accurate and simple, in both the microfabrication and the heat transfer model. However, in most cases we have to sacrifice one to pursue the other (Fig. 1.1). Some representative techniques are depicted as examples to illustrate this point, and summarized in Table 1.1, which is a high-level guideline to selecting an appropriate technique.

The remainder of Chapter 1 is devoted to a brief introduction of each of these techniques, deferring the detailed technical discussion to Chapter 2 (corresponding section numbers are given in Table 1.1).

1.2.   Resistance thermometry

The techniques presented throughout this book rely heavily on resistance thermometry, the underlying physics of which is the

Table 1.1   A high-level guideline to selecting an electrothermal measurement technique.

	Bulk (iso)	Bulk (aniso)	Thin film (⊥)	Thin film (∥)	Nanotube/ wire	Bio-tissue
3-omega (1.3)	2.2.1; 3.3, 3.5; Appendices B and D	2.2.2	2.3.1	2.3.2(E)	2.4(A)	2.5
Suspend. μ-fab device (1.4)				2.3.2(A)		
Self-heating (1.5)				2.3.2(B)		
T-bridge (1.6)				2.3.2(C)		
Central-line heater (1.7)				2.3.2(D)		
Heat spreader (1.8)				2.3.2(F); 3.2, 3.4, 3.5; 4.3, 4.4		

The columns give the physical character and morphology of the sample, while the rows correspond to six popular techniques, each of which is depicted in Fig. 1.1 and briefly introduced in the corresponding section of Chapter 1 indicated in parentheses. Entries in the table specify the relevant sections of this book. For example, the central-line heater method is suitable for measuring the in-plane thermal conductivity of thin films, as discussed further in Sec. 2.3.2(D). For nanotubes/wires and in-plane films, samples generally should be suspended above any substrate, with the exception of the heat spreader method. Abbreviations: iso = isotropic; aniso = anisotropic; ⊥ = cross-plane; ∥ = in-plane; X = not recommended.

temperature-dependent electrical resistance $R_e(T)$ of a metal. We focus on small temperature excursions such that the $R_e(T)$ curve can be linearized as

$$R_e = R_{e,0}[1 + \alpha(T - T_0)], \qquad (1.1)$$

where $R_{e,0}$ and R_e are electrical resistances of the thermometer at a reference temperature T_0 and the temperature of interest T, respectively, and α is the temperature coefficient of resistance defined as $\alpha = (dR/dT)/R_{e,0}$ and evaluated at T_0. An example of the uncertainty analysis of an $R_e(T)$ calibration is given in Sec. 4.4.2, and readers interested in additional information about resistance thermometry are referred to [3].

1.3. 3ω method

The 3ω method is arguably the most widely used electrothermal measurement technique, because of its adaptability to measure various structures (see Table 1.1), its relatively straightforward microfabrication, and its simple heat transfer model. This method was developed by Cahill [4] in the late 1980s to measure the thermal conductivity of bulk samples (as in Fig. 1.2) and films. As an indication of the impact of such microscale techniques on the broader

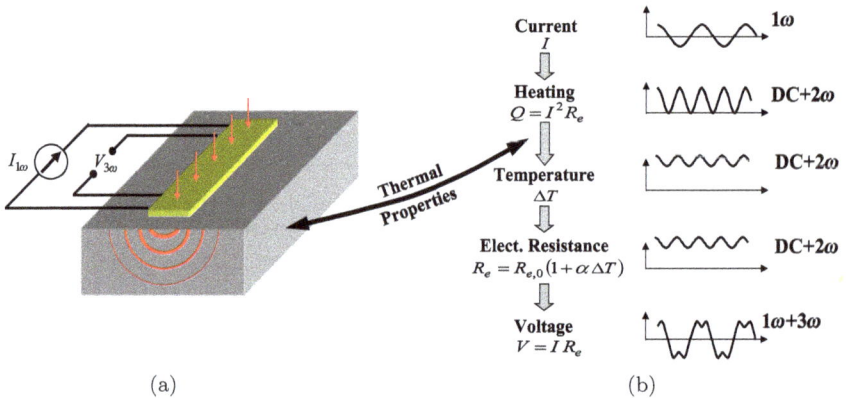

(a)   (b)

Fig. 1.2   Schematic of the classic 3ω method to measure the thermal conductivity of a substrate. (a) A microfabricated metal line (yellow) on top of the sample (gray) serves simultaneously as heater and thermometer. (b, from top to bottom) An AC current with frequency of 1ω drives the experiment and causes Joule heating with frequencies of 0ω (DC) and 2ω. The Joule heating in turn results in a 2ω temperature oscillation superposed with a DC offset. This temperature response, which depends on the thermal properties of the sample, increases the electrical resistance in direct proportion. Finally the 1ω driving current and the resistance response (at DC and 2ω) produce voltages at 1ω and 3ω, both of which contain information of the thermal property of the sample. In practice it is the third harmonic which is most useful, giving this method its name.

heat transfer community, the 3ω method is now used as an example problem in a celebrated undergraduate textbook (see [5, p. 300]).

The basic idea of the classic 3ω method [4] is illustrated in Fig. 1.2. The measurement is driven by an AC current with angular frequency ω through a heater line microfabricated on the sample. The resulting Joule heating oscillates at a frequency of 2ω, which correspondingly produce a temperature field oscillating at the same frequency. As a response to this temperature oscillation, the electrical resistance of the heater line also carries an AC component with a frequency of 2ω, superposed on a DC component. Finally, combining these two components of the electrical resistance with the driving AC current, we obtain two superposed voltage components, $V_{1\omega}$ and $V_{3\omega}$, both of which contain important information of thermal properties of the sample. In particular, the 3ω component is widely used, and known as the 3ω method.

A detailed discussion of the technical aspects of this classic 3ω method is given in Sec. 2.2.1. We also give an example in Sec. 3.3 on how to estimate some key electrical parameters prior to the experiment, and another example in Sec. 3.5 on a sanity check to confirm the measurement is dominated by the thermal signal rather than various non-thermal artifacts.

The classic 3ω method for measuring isotropic bulk samples has also been extended to measure the thermal properties of anisotropic bulk samples (Sec. 2.2.2), thin films along both cross-plane (Sec. 2.3.1) and in-plane (Sec. 2.3.2(E)) directions, nanotubes and nanowires (Sec. 2.4(A)), and even liquids, biological tissues, and other soft matter (Sec. 2.5).

Example 1.1. Beginning with a driving AC current passing through the heater line, $I = I_{1\omega}\sin(\omega t)$, where ω is the angular frequency of the driving current, obtain expressions for each of the steps in the flowchart of Fig. 1.2(b). Assuming that the temperature response (ΔT) is related to the Joule heating (Q) by $\Delta T = Q \cdot R$, where R represents an effective thermal resistance of the sample underneath the heater line (Fig. 1.2(a)).

Solution: With the driving AC current, Joule heating (to leading order) is generated as

$$Q = I^2 R_{e,0} = I_{1\omega}^2 R_{e,0} \sin^2(\omega t) = I_{1\omega}^2 R_{e,0} \left[\frac{1}{2} - \frac{1}{2} \cos(2\omega t) \right],$$

where $R_{e,0}$ is the leading term of the electrical resistance of the heater line. Accumulation of the heat causes temperature response:

$$\Delta T = Q \cdot R = I_{1\omega}^2 R_{e,0} R \left[\frac{1}{2} - \frac{1}{2} \cos(2\omega t) \right].$$

This temperature variation leads to an electrical resistance response:

$$R_e = R_{e,0}(1 + \alpha \Delta T) = R_{e,0} \left\{ 1 + \alpha I_{1\omega}^2 R_{e,0} R \left[\frac{1}{2} - \frac{1}{2} \cos(2\omega t) \right] \right\},$$

and a corresponding voltage response:

$$V = I \cdot R_e = [I_{1\omega} \sin(\omega t)] R_{e,0} \left\{ 1 + \alpha I_{1\omega}^2 R_{e,0} R \left[\frac{1}{2} - \frac{1}{2} \cos(2\omega t) \right] \right\}.$$

Applying a trigonometric identity, we arrive at

$$V = \left[I_{1\omega} R_{e,0} + \frac{3}{4} \alpha I_{1\omega}^3 R_{e,0}^2 R \right] \sin(\omega t) - \left[\frac{1}{4} \alpha I_{1\omega}^3 R_{e,0}^2 R \right] \sin(3\omega t).$$

Thus, the 1ω component voltage can be expressed as

$$V_{1\omega} = \left[I_{1\omega} R_{e,0} + \frac{3}{4} \alpha I_{1\omega}^3 R_{e,0}^2 R \right] \sin(\omega t),$$

and likewise the 3ω component is

$$V_{3\omega} = - \left[\frac{1}{4} \alpha I_{1\omega}^3 R_{e,0}^2 R \right] \sin(3\omega t).$$

1.4. Suspended microfabricated-device method

The suspended microfabricated device method is another signature work of thermal measurement at the nanoscale. This method achieves outstanding simplicity in its heat transfer model, namely, one-dimensional (1D) steady-state heat transfer, although it requires complex microfabrication. This method was developed in the early 2000s by Shi, Kim, Li, and coworkers [7, 8], who applied it for a number of pioneering works including the first measurement of

all five leads connect to T_∞

V^+_{Htr} V^-_{Htr} I^-_{Htr} V^-_{TE} I^+_{Sens}

Heating island

Sample

Sensing island

I^+_{Htr} V^+_{TE} I^-_{Sens} V^-_{Sens} V^+_{Sens}

all five leads connect to T_∞

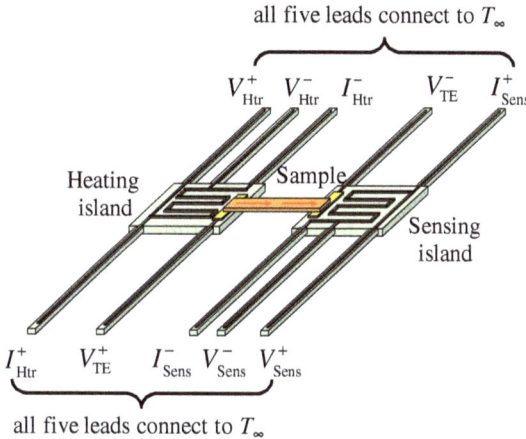

Fig. 1.3 Schematic of the suspended microfabricated device method, redrawn after [6, 7]. Two islands are microfabricated for heating and sensing, respectively. Each island is patterned with a serpentine heater/thermometer (platinum), and supported by five long SiN_x beams with electrodes, four of which are for four-probe measurement of the heater/thermometer with the fifth for thermoelectric measurement of the sample. The heating and sensing islands are thermally coupled by the sample, which undergoes 1D steady-state heat transfer.

the thermal conductance of individual multiwall carbon nanotubes [9], silicon nanowires [1], and silicon/silicon-germanium superlattice nanowires [10], and is another example problem of the celebrated undergraduate textbook (see [5, p. 110]).

As shown in Fig. 1.3, in a typical implementation two isolated islands are microfabricated from silicon nitride (SiN_x) and patterned with a serpentine platinum thin film for heating and temperature sensing. A sample bridges the two islands, allowing 1D heat transfer through it. Each island is supported by five long SiN_x beams with Pt electrical leads. Four leads are used for four-probe resistance measurement of the Pt serpentine, and the fifth lead enables electrical and thermoelectric probing of the sample itself. By measuring the Joule heating at the heating island and the temperature of the two islands by resistance thermometry, the thermal conductance of the sample is obtained.

More technical details of this method will be discussed in Sec. 2.3.2(A).

1.5.   Distributed self-heating method

The distributed self-heating method is appealing for its simple experimental configuration, requiring no data acquisition beyond the familiar *I–V* curve of the sample. The basic idea is illustrated in Fig. 1.4. The sample also serves as the heater and the temperature sensor. A DC or AC current passes through a suspended, conducting sample, and generates Joule heating uniformly along the sample. Knowing the Joule heating power and the average temperature of the sample (from resistance thermometry of the sample itself), the thermal conductivity of the sample is extracted from the 1D steady-state heat diffusion equation with uniform heat generation. This technique has been used to study the thermal properties of carbon nanotubes (CNTs) [11–13] and microwires [14, 15].

This method involves a number of thermal subtleties which must also be considered to ensure an accurate measurement, such as the thermal and electrical contact resistances where the film edge meets the substrate, the substrate spreading resistance, the radiation losses from the film surfaces, and the linearity of the *I–V* curve. More

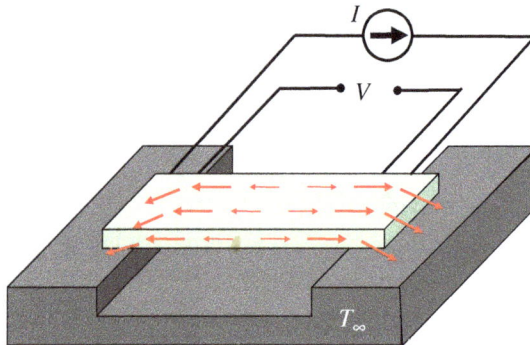

Fig. 1.4   Schematic of the distributed self-heating method. The sample, suspended above a trench, also serves as its own Joule heater and distributed thermometer. The thermal conductivity of the sample is obtained by analyzing the measured *I–V* curve with a 1D steady-state heat diffusion equation with a volumetric heat source.

discussion about the technical aspects of this self-heating method will be given in Sec. 2.3.2(B).

Example 1.2. For steady-state Joule heating of the sample depicted in Fig. 1.4, obtain the parabolic temperature profile $T(x)$ along the sample, where x is the direction spanning the trench. Also calculate the average temperature of the sample, $\langle T \rangle$. Focus on the ideal scenario which clamps the temperature of the two ends of the sample to T_∞, thus neglecting the contact and substrate spreading resistance. Similarly, neglect radiation and convection losses from the sample surfaces. The sample has length L and cross-sectional area A_{cr}.

Solution: The governing equation is

$$\frac{d^2 T}{dx^2} + \frac{Q'''}{k} = 0,$$

where

$$Q''' = (IV)/(A_c L)$$

is the volumetric heat generation due to Joule heating. The boundary conditions are

$$T\left(x = \frac{L}{2}\right) = T_\infty,$$

$$\left.\frac{dT}{dx}\right|_{x=0} = 0,$$

where we measure x from the symmetry plane. Solving, we obtain the temperature profile:

$$T(x) = T_\infty - \frac{IV}{2k A_c L} x^2 + \frac{IVL}{8k A_c}.$$

In real experiments using resistance thermometry, we measure the average temperature across the sample:

$$\langle T \rangle = \frac{1}{L} \int_{-L/2}^{L/2} T(x) dx$$

$$= T_\infty + \frac{IVL}{12k A_c}.$$

1.6.   Variations of the distributed self-heating method, including T-bridge method

One variation of the distributed self-heating method is to omit the trench underneath the sample, thus allowing thermal contact between the entire sample and the substrate [16]. This variation simplifies the microfabrication, with the corresponding tradeoff of complicating the heat transfer model. A second variation is to allow an electrically nonconducting sample, but a metal coating of the sample is required [17]. A third variation is the so-called T-bridge method [18, 19]. As shown in Fig. 1.5, in this variation the sample is liberated from the duty of heater and sensor, and serves solely as a shunting thermal resistor. More technical details will be discussed in Sec. 2.3.2(C).

Example 1.3. Show how the $T(x)$ temperature profile along the heater/sensor of the T-bridge method differs from the parabolic profile obtained in Example 1.2. For simplicity, here we neglect the width of the sample as compared to the length of the heater/sensor. Use similar idealizations as in Example 1.2 with regards to perfect thermal contact and neglecting surface heat losses, from both

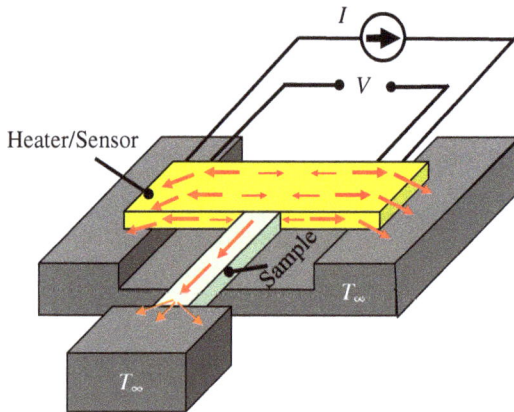

Fig. 1.5 Schematic of the T-bridge method. This method can be viewed as a variation of the distributed self-heating method. The sample (green) acts as a shunting thermal resistor in parallel with the heater/sensor (yellow).

heater/sensor and sample. The heater/sensor has length L and cross-sectional area A_{cr}. Express your result as a function of the dimensionless ratio $\gamma = R_{Htr}/(4R_{sampl})$, where $R_{Htr} = L/(k_{Htr}A_{cr})$ is the thermal resistance of the heater/sensor, and R_{sampl} is the thermal resistance of the sample.

Solution: The governing equation applying to the heater/sensor (gold in Fig. 1.5) is the same as the distributed self-heating method in Example 1.2:

$$\frac{d^2T}{dx^2} + \frac{Q'''}{k} = 0,$$

where

$$Q''' = (I \cdot V)/(A_{cr} \cdot L)$$

is the volumetric heat generation due to Joule heating.

The first boundary condition on the end of the heater/sensor is also the same as the distributed self-heating method:

$$T\left(x = \frac{L}{2}\right) = T_\infty.$$

The key difference comes from the boundary condition in the middle of the heater/sensor, where now heat is leaking through the sample:

$$kA_{cr}\left(\left.\frac{dT}{dx}\right|_{x=0+} - \left.\frac{dT}{dx}\right|_{x=0-}\right) = \frac{T(0) - T_\infty}{R_{sampl}}.$$

Solving, we obtain the temperature profile, which is modified from a parabola:

$$T(x) = T_\infty + \frac{Q}{8}R_{Htr}\left[-\left(\frac{x}{L/2}\right)^2 + \frac{\gamma}{1+\gamma}\left|\frac{x}{L/2}\right| + \frac{1}{1+\gamma}\right],$$

where $Q = Q''' \cdot (A_{cr} \cdot L)$ is the Joule heating applied to the heater/sensor. The dimensionless temperature profiles for five different γ are shown in Fig. 1.6.

As a sanity check, in the limit $\gamma = R_{Htr}/(4R_{sampl}) = 0$, the thermal path through the sample is broken. Since there is no heat flowing through the sample, the solution recovers the parabolic

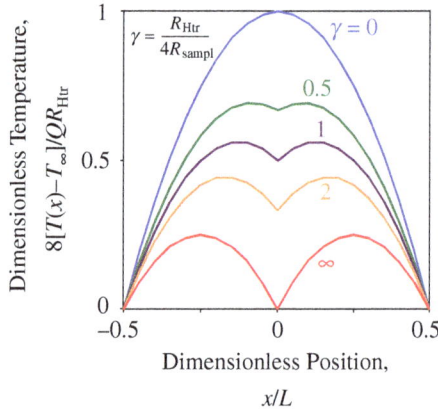

Fig. 1.6   Dimensionless temperature profile of the T-bridge method for different values of γ, the ratio between the thermal resistance of the wire heater and the sample.

temperature profile of Example 1.2 as expected, and shown as the blue line in Fig. 1.6.

On the other hand, in limit $\gamma = \infty$, the sample is a thermal short circuit, so that the temperature at the midpoint of the wire heater is clamped to T_∞. This results in two smaller parabolas as shown in red in Fig. 1.6, each corresponding to Example 1.2 with $L \to L/2$ and $V \to V/2$.

It can be proven that the sensitivity of this technique is the best for intermediate γ in the vicinity of unity [19]. Recently, a generalization to samples with a finite width has also been developed [20].

1.7.   Central-line heater method

The central-line heater method [21–25], sometimes called a Völklein method [26], is another technique with an appealingly simple heat transfer model, namely, 1D steady-state heat transfer. As shown in Fig. 1.7, the placement of the heater line in the middle of the suspended sample takes advantage of symmetry. In this case, the Joule heating separates equally into two halves, each flowing to the corresponding end of the sample.

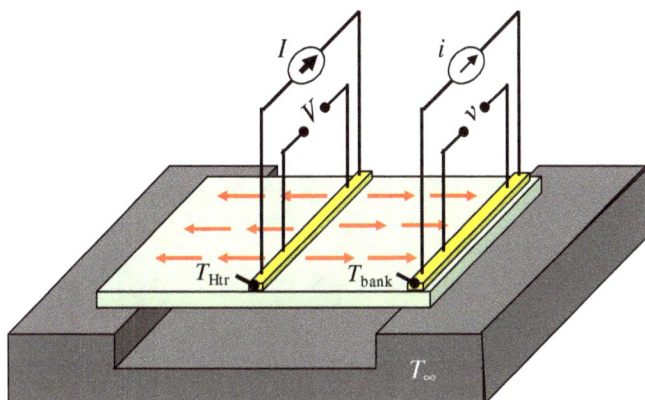

Fig. 1.7 Schematic of the central-line heater method (sometimes called a Völklein method). This technique takes advantage of the symmetry of the structure, resulting in a 1D steady-state heat transfer on both sides of the central heater line. Temperatures are measured at both metal lines. Care must be taken about issues such as thermal contact and heat spreading from the edge of the sample to the substrate, as well as radiation losses from the sample.

As compared to the suspended microfabricated-device method (Sec. 1.4), the central-line heater method further simplifies the heat transfer model by eliminating the heat losses through the five long SiN_x beams of the heating island. However, aligning the voltage probes requires some care. Another distinction from the microfabrication perspective is that the central-line heater method requires a relative large sample area to place the heater and sensor, while the suspended microfabricated-device method does not have this constraint. More detailed technical discussion will be found in Sec. 2.3.2(D).

1.8. Heat spreader method

The heat spreader method [27, 28] can be viewed as a variation of the central-line heater method. As compared to the central-line heater method (Fig. 1.7), the heat spreader method avoids the challenge of suspending a sample across a trench, which significantly simplifies the microfabrication. As a tradeoff, the heat transfer model of the heat spreader method is more complicated.

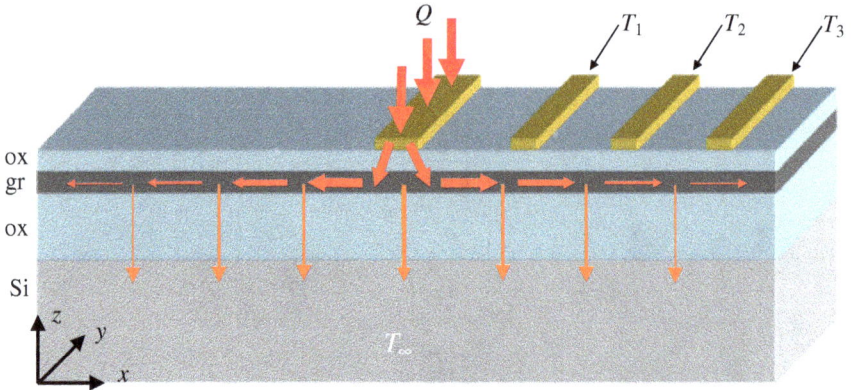

Fig. 1.8   Schematic of a heat spreader method used to measure k for graphene (gr) encased between top and bottom SiO_2 films (ox), on a high-k Si wafer which acts like a heat sink. Four metal lines (gold color) are patterned on top of the sample, to act as a heater and three resistive thermometers. The key physics determining the temperature response are the lateral heat spreading along the encased graphene flake and the vertical leakage through the lower SiO_2 film. k for graphene is obtained by fitting the measured temperature profile to a heat transfer model.

Figure 1.8 shows a schematic of the heat spreader method developed by Jang, Chen, *et al.* [28] to measure the in-plane k of encased graphene. The graphene thin film is encased between top and bottom oxide layers, a configuration which is relevant to microelectronics applications, where graphene might be used as transistors, interconnects, and/or thermal management materials, likely surrounded by dielectric isolation. The silicon substrate acts as a heat sink. A heater and three temperature sensors are microfabricated on top of the upper oxide layer, which is used to electrically isolate these electrodes from the graphene. The heater power Q_{Htr} flows vertically through the stack into the Si heat sink, while simultaneously spreading laterally through the high-k graphene layer. Compared to a control experiment with no graphene layer, the configuration in Fig. 1.8 results in higher temperatures at the sensors $T_1 - T_3$. Finally, the in-plane k of the graphene layer is inferred by fitting the three measured temperatures (T_1, T_2, and T_3) to a one-parameter thermal model.

Example 1.4. The simplest thermal model for analyzing Fig. 1.8 is to treat the graphene heat spreader as a fin, where the effective "convection" coefficient h_{eff} represents the vertical conduction through the lower SiO_2 layer into the Si heat sink. Thus $h_{\text{eff}} = k_{\text{BotOx}}/t_{\text{BotOx}}$, where k_{BotOx} and t_{BotOx} are the thermal conductivity and thickness of the lower SiO_2 layer. Use this model to obtain an exponentially-decaying temperature profile $T(x)$ along the graphene heat spreader. You may assume the sample is very large in the $\pm x$ direction, and that there is no variation out of the plane in y, making this a two-dimensional problem.

Solution: The standard governing equation for a fin is

$$\frac{d^2T}{dx^2} - \frac{h_{\text{eff}} P}{k A_{\text{cr}}}(T - T_\infty) = 0,$$

where k, P, and A_{cr} are the thermal conductivity, perimeter, and cross-sectional area of the fin. Corresponding to Fig. 1.8, $k_{\text{gr},\parallel}$ and P are the in-plane thermal conductivity and the "wetted perimeter" of the graphene flake. For a flake of width w along the y direction and thickness t_{gr}, we have $A_{\text{cr}} = w \cdot t_{\text{gr}}$ and $P = w$ because the flake only conducts heat out through its lower face.

We take the first boundary condition to be a prescribed temperature

$$T(x = 0) = T_b.$$

This could be obtained by measuring the electrical resistance of the heater (Fig. 1.8), although this is not actually necessary for this example problem because we care only about the shape of the temperature profile.

The second boundary condition is for the end of the fin, which might be a prescribed flux or a temperature. Here for simplicity we assume the graphene flake is infinitely long, and thus in the "long fin limit" the second boundary conduction becomes

$$T(x \to \infty) = T_\infty.$$

Solving, we obtain the temperature profile:

$$T(x) = T_\infty + (T_b - T_\infty)\exp(-\beta x),$$

where $\beta^{-1} = \sqrt{k_{\mathrm{gr},\parallel} A_{\mathrm{cr}}/h_{\mathrm{eff}} P}$ is a characteristic decay length. For the configuration of Fig. 1.8, we have $\beta^{-1} = \sqrt{(k_{\mathrm{gr},\parallel}/k_{\mathrm{BotOx}}) \cdot t_{\mathrm{gr}} \cdot t_{\mathrm{BotOx}}}$. Thus, by measuring the shape of the temperature profile, it is possible to determine β^{-1} and $k_{\mathrm{gr},\parallel}$ of the graphene flake.

It turns out that real experiments fall well outside of this idealized fin regime, requiring a more sophisticated analysis as discussed further in Sec. 2.3.2(F). In addition, discussions of the thermal design, control experiments, and sanity checks for this method can be found in Secs. 3.2, 3.4, and 3.5; and the uncertainty and sensitivity analysis in Secs. 4.3 and 4.4.

1.9.   Closing remarks

In this chapter we have briefly introduced the basic ideas and physical pictures of six common techniques for electrothermal measurements at the nanoscale. We will revisit each of these in more depth in subsequent chapters, as referenced at the end of each preceding section and summarized in Table 1.1.

References

[1] D. Li, Y. Wu, P. Kim, L. Shi, P. Yang and A. Majumdar, "Thermal conductivity of individual silicon nanowires," *Appl. Phys. Lett.* **83**(14), pp. 2934–2936, 2003.

[2] Z. Wang, J. E. Alaniz, W. Jang, J. E. Garay and C. Dames, "Thermal conductivity of nanocrystalline silicon: Importance of grain size and frequency-dependent mean free paths," *Nano Lett.* **11**(6), pp. 2206–2213, 2011.

[3] C. Dames, "Resistance Temperature Detectors," in *Encyclopedia of Micro- and Nanofluidics*, ed. D. Li. Springer, 2008.

[4] D. G. Cahill, "Thermal conductivity measurement from 30 to 750 K: The 3-omega method," *Rev. Sci. Instrum.* **61**(2), pp. 802–808, 1990.

[5] F. P. Incropera, D. P. DeWitt, T. L. Bergman and A. S. Lavine, *Fundamentals of Heat and Mass Transfer*, 6th ed. Wiley, 2007.

[6] L. Shi, D. Y. Li, C. H. Yu, W. Y. Jang, D. Kim, Z. Yao, P. Kim and A. Majumdar, "Measuring thermal and thermoelectric properties of one-dimensional nanostructures using a microfabricated device," *J. Heat Transfer* **125**(5), pp. 881–888, 2003.

[7] D. Li, "Thermal transport in individual nanowires and nanotubes," PhD dissertation, University of California at Berkeley, 2002.

[8] L. Shi, "Mesoscopic thermophysical measurements of microstructures and carbon nanotubes," PhD dissertation, University of California at Berkeley, 2001.

[9] P. Kim, L. Shi, A. Majumdar and P. McEuen, "Thermal transport measurements of individual multiwalled nanotubes," *Phys. Rev. Lett.* **87**(21), p. 215502, 2001.

[10] D. Li, Y. Wu, R. Fan, P. Yang and A. Majumdar, "Thermal conductivity of Si/SiGe superlattice nanowires," *Appl. Phys. Lett.* **83**, p. 3186, 2003.

[11] H. Y. Chiu, V. V. Deshpande, H. W. C. Postma, C. N. Lau, C. Mikó, L. Forró and M. Bockrath, "Ballistic phonon thermal transport in multiwalled carbon nanotubes," *Phys. Rev. Lett.* **95**(22), pp. 1–4, 2005.

[12] E. Pop, D. Mann, J. Cao, Q. Wang, K. Goodson and H. Dai, "Negative differential conductance and hot phonons in suspended nanotube molecular wires," *Phys. Rev. Lett.* **95**(15), pp. 1–4, 2005.

[13] L. Lu, W. Yi and D. L. Zhang, "3ω method for specific heat and thermal conductivity measurements," *Rev. Sci. Instrum.* **72**(7), p. 2996, 2001.

[14] A. Potts, M. J. Kelly, D. G. Hasko, J. R. A. Cleaver, H. Ahmed, D. A. Ritchie, J. E. F. Frost and G. A. C. Jones, "Lattice heating of free-standing ultra-fine GaAs wires by hot electrons," *Semicond. Sci. Technol.* **7**, pp. 8231–8234, 1992.

[15] K. Yoh, A. Nishida and H. Kawahara, "Electron and thermal transport in InAs single-crystal free-standing wires," *Semicond. Sci. Technol.* **9**, pp. 961–965, 1994.

[16] E. Pop, D. A. Mann, K. E. Goodson and H. Dai, "Electrical and thermal transport in metallic single-wall carbon nanotubes on insulating substrates," *J. Appl. Phys.* **101**(9), p. 93710, 2007.

[17] W. Liu and M. Asheghi, "Thermal conductivity measurements of ultra-thin single crystal silicon layers," *J. Heat Transfer* **128**, pp. 75–83, 2006.

[18] M. Fujii, X. Zhang, H. Xie, H. Ago, K. Takahashi, T. Ikuta, H. Abe and T. Shimizu, "Measuring the thermal conductivity of a single carbon nanotube," *Phys. Rev. Lett.* **95**, p. 65502, 2005.

[19] C. Dames, S. Chen, C. T. Harris, J. Y. Huang, Z. F. Ren, M. S. Dresselhaus and G. Chen, "A hot-wire probe for thermal measurements of nanowires and nanotubes inside a transmission electron microscope," *Rev. Sci. Instrum.* **78**(10), p. 104903, 2007.

[20] J. Kim, D.-J. Seo, H. Park, H. Kim, H.-J. Choi and W. Kim, "Extension of the T-bridge method for measuring the thermal conductivity of two-dimensional materials, **88**, p. 54902, 2017.

[21] Y. C. Tai, C. H. Mastrangelo and R. S. Muller, "Thermal conductivity of heavily doped low-pressure chemical vapor deposited polycrystamne silicon films," *J. Appl. Phys.* **63**, pp. 1442–1447, 1988.

[22] J. E. Graebner, J. A. Mucha, L. Seibles and G. W. Kammlott, "The thermal conductivity on silicon of chemical-vapor-deposited diamond films," *J. Appl. Phys.* **71**, pp. 3143–3146, 1992.

[23] P. G. Sverdrup, S. Sinha, M. Asheghi, S. Uma and K. E. Goodson, "Measurement of ballistic phonon conduction near hotspots in silicon," *Appl. Phys. Lett.* **78**(21), pp. 3331–3333, 2001.

[24] M. Asheghi, K. Kurabayashi, R. Kasnavi and K. E. Goodson, "Thermal conduction in doped single-crystal silicon films," *J. Appl. Phys.* **91**(8), pp. 5079–5088, 2002.

[25] D. Song and G. Chen, "Thermal conductivity of periodic microporous silicon films," *Appl. Phys. Lett.* **84**(5), pp. 687–689, 2004.

[26] F. Völklein, "Thermal conductivity and diffusivity of a thin film SiO_2–Si_3N_4 sandwich system," *Thin Solid Films* **188**, pp. 27–33, 1990.

[27] M. Asheghi, M. N. Touzelbaev, K. E. Goodson, Y. K. Leung and S. S. Wong, "Temperature-dependent thermal conductivity of single-crystal silicon layers in SOI substrates," *J. Heat Transfer* **120**, p. 30, 1998.

[28] W. Jang, Z. Chen, W. Bao, C. N. Lau and C. Dames, "Thickness-dependent thermal conductivity of encased graphene and ultrathin graphite," *Nano Lett.* **10**(10), pp. 3909–3913, 2010.

Chapter 2

Which Tool Should You Choose?

"Water shapes its course according to the nature of the ground over which it flows; the soldier works out his victory in relation to the foe whom he is facing. Therefore, just as water retains no constant shape, so in warfare there are no constant conditions. He who can modify his tactics in relation to his opponent and thereby succeed in winning, may be called a heaven-born captain."

— Sun Tzu, The Art of War

2.1. Introduction

In this chapter, we discuss how to choose a suitable measurement technique from the toolbox outlined in Chapter 1. The criterion is mainly based on the sample's dimensionality and geometry, e.g., bulk, film, or nanotube/wire (also referred to as 3D, 2D, or 1D, respectively), and this chapter is organized accordingly. In some cases more than one technique is applicable to a specific sample type, so trade-offs are also discussed.

Most of the techniques presented below have been widely discussed in previous literature, so here we take a high-level approach, minimizing mathematical details to emphasize the most useful final expressions for the thermal properties of interest. We also present some more recent developments in measuring anisotropic solids (Sec. 2.2.2) and biological tissues (Sec. 2.5), which have not yet been reviewed elsewhere.

We also seek to convey the message that behind any experimental technique there is a corresponding heat transfer model, and any model relies on various assumptions which must be justified in the real experiments. Such restrictions often lead to substantial constraints on the practical experimental parameters, as discussed throughout this chapter. One good example is the process of selecting the proper frequency range for the classic 3ω method, as discussed in Sec. 2.2.1.

2.2. Bulk samples

Although this is a book on measurements at the nanoscale, in most cases understanding the thermal transport physics of a nanostructure also requires knowledge of the thermal properties of its bulk form. Here we present well-established 3ω methods for solid substrates (semi-infinite thickness) with both isotropic and anisotropic thermal properties, which also form the basis for several modified 3ω methods presented later for nanostructure measurements.

2.2.1. *Isotropic solids: The classic 3ω method for a semi-infinite substrate*

The classic 3ω method is widely used for measuring the thermal conductivity of bulk samples, especially those with low thermal conductivities. As compared to traditional steady-state techniques, the 3ω method has a faster measurement time, is relatively immune from radiation losses, and as an AC method is insensitive to DC artifacts from parasitic thermoelectric voltages [1, 2].

The most important heat transfer concept in the 3ω method is the thermal penetration depth, defined as

$$L_p = \sqrt{\frac{D_\text{sub}}{\omega_H}}, \tag{2.1}$$

where D_sub is the thermal diffusivity of the substrate and ω_H is the heating frequency (to be distinguished from the frequency of the driving current, ω, with $\omega_H = 2\omega$). Intuitively, L_p is a characteristic length scale that describes how far the heat can penetrate through the sample from the heater. As indicated in Eq. (2.1) and visualized in Fig. 2.1, a powerful feature of the periodic heating problem is that the penetration depth can be tuned by varying the heating frequency:

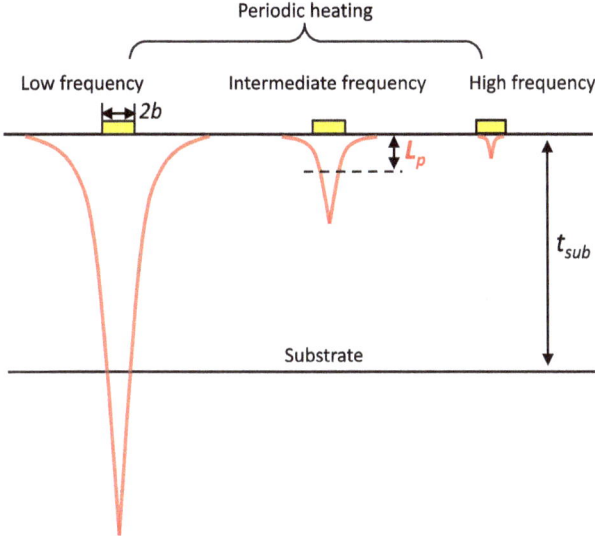

Fig. 2.1 The concept of penetration depth $[L_p = (D_{\text{sub}}/\omega_H)^{0.5}]$ in a periodic heating problem. The red line represents the amplitude envelope of the oscillating thermal wave which penetrates from the heater line of width $2b$ (gold color) into the substrate. The key feature is that L_p can be tuned by varying the heating frequency ω_H. The intermediate scenario satisfies the basic assumptions of the classic 3ω method, i.e., $b \ll L_p \ll t_{\text{sub}}$.

the higher the heating frequency, the shorter the penetration depth. This feature offers an approach to localize heat.

Using this concept of the penetration depth, we discuss the key assumptions of the heat transfer model of the classic 3ω method, which requires a very narrow, infinitely long, heater line which periodically heats a semi-infinite substrate [1]. Although infinities are lacking in real experiments, these idealizations are still a good approximation as long as the experiment lies in the intermediate regime depicted in Fig. 2.1:

(1) $L_p \gg b$, where b is the half-width of the heater line;
(2) $L_p \ll l$, where l is the length of the heater line (into the page);
(3) $L_p \ll t_{\text{sub}}$, where t_{sub} is the thickness of the substrate.

The left scenario in Fig. 2.1 is inconvenient for analysis because the heating frequency is so low as to violate the third requirement;

likewise, in the right scenario the heating frequency is so high that
the first requirement is violated. Thus, these two requirements set
upper and lower bounds for the heating frequency of 3ω method.
(The second requirement is rarely limiting in practice, because l
is determined by microfabrication and one nearly always can use
$l \gg t_{\text{sub}}$.)

Regarding instrumentation, a fundamental challenge is the fact
that the 3ω voltages of interest are typically 100–1000 times smaller
than the inherent 1ω background. This is most commonly addressed
by using some variation of an adjustable resistor and a subtraction
circuit, for example as presented in the original 1990 paper (see
[1, Fig. 4]). Figure 2.2 shows an alternative, which forgoes the
background subtraction circuit but requires careful attention to the
dynamic reserve settings to avoid overloading the lock-in amplifier
[3]. For the AC current source, options include a homemade V-to-I
converter (see an example in Appendix D), a turnkey commercial
source as depicted in Fig. 2.2, or, in special circumstances, an AC
voltage source in series with a carefully-chosen ballast resistor [3].

Figure 2.3 shows typical raw data for the third harmonic volt-
ages in the preferred intermediate frequency regime $[b \ll L_p \ll \min(t_{\text{sub}}, l)]$. Both the in-phase and out-of-phase signals include

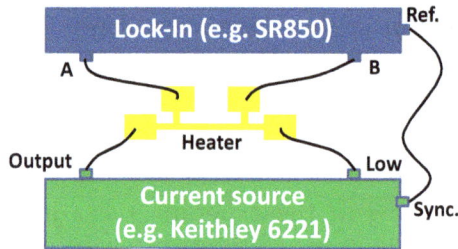

Fig. 2.2   One possible instrumentation setup for the 3ω method. The heater
deposited on top of the sample is driven by an AC current source, and its voltage
drop monitored by a lock-in amplifier. This is a typical configuration for 4-probe
resistance thermometry. Note that although the original setup with a cancellation
resistor and a subtraction circuit (see [1, Fig. 4]) is still most common, the simpler
setup here is also possible if the lock-in has enough dynamic reserve and it is
carefully configured.

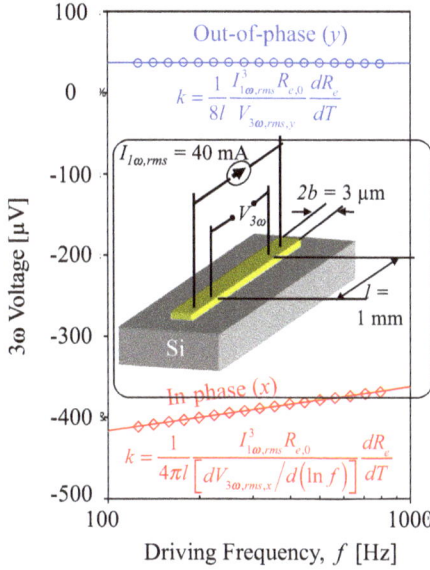

Fig. 2.3 A typical 3ω measurement performed on an undoped Si wafer at 310 K. The in-phase (red symbols) and out-of-phase (blue symbols) 3ω voltages are plotted as a function of the driving frequency $f = \omega/2\pi$ (log scale). The red line is a linear fit to the in-phase 3ω voltage, the slope of which gives the thermal conductivity of 137.8 W/m·K using the equation in red; similarly, the blue horizontal line is the constant value fit to the out-of-phase 3ω voltage, which gives the thermal conductivity of 140.3 W/m·K using the equation in blue. As depicted in the inset schematic, the driving current in this measurement is 40 mA (rms), and the heater line is 3 μm wide and 1 mm long (between the two inner voltage probes).

information about the thermal conductivity of the bulk sample. From the in-phase 3ω voltage, the thermal conductivity can be obtained as [1, 3]

$$k = \frac{1}{4\pi l} \frac{I_{1\omega,\text{rms}}^3 R_{e,0}}{[dV_{3\omega,\text{rms},x}/d(\ln f)]} \frac{dR_e}{dT}, \tag{2.2}$$

where l and $R_{e,0}$ are the length and the electrical resistance (at the ambient reference temperature T_0) of the heater line between the two voltage probes, dR_e/dT is the slope of the electrical resistance of the heater line as a function of the temperature, $I_{1\omega,\text{rms}}$ is the root-mean-square (rms) 1ω driving current at a frequency $f = \omega/2\pi$, $V_{3\omega,\text{rms},x}$

is the in-phase component (subscript x) of the rms 3ω voltage, and correspondingly $dV_{3\omega,\mathrm{rms},x}/d(\ln f)$ is the slope of $V_{3\omega,\mathrm{rms},x}$ as a function of natural logarithm of f.

Note that Eq. (2.2) takes advantage of measurements spanning a range of frequencies, rather than simply a single point measurement. Similarly, to incorporate information from a range of $I_{1\omega,\mathrm{rms}}$, the expression can be further generalized to

$$k = \frac{1}{4\pi l} \frac{R_{e,0}}{\left[\dfrac{\partial^2 V_{3\omega,\mathrm{rms},x}}{\partial(\ln f)\partial(I^3_{1\omega,\mathrm{rms}})}\right]} \frac{dR_e}{dT}. \tag{2.3}$$

Although such a $(f, I_{1\omega,\mathrm{rms}})$ sweep is rarely part of routine practice, when setting up a new experiment we find that verifying this universal $V_{3\omega,\mathrm{rms},x} \propto I^3_{1\omega,\mathrm{rms}}$ scaling is often a worthwhile diagnostic check (see an example in Fig. 3.8). We highly recommended it for researchers setting up 3ω measurements for the first time.

As another check the out-of-phase component (subscript y) could also be used to independently extract the substrate thermal conductivity as

$$k = \frac{1}{8l} \frac{I^3_{1\omega,\mathrm{rms}}}{V_{3\omega,\mathrm{rms},y}} \frac{R_{e,0}}{dT} \frac{dR_e}{dT}, \tag{2.4}$$

where $V_{3\omega,\mathrm{rms},y}$ is the out-of-phase component of the rms value of the 3ω voltage. This out-of-phase approach is generally considered not as robust as the in-phase approach of Eq. (2.2), but it still can be a helpful check especially for novices and when setting up a new measurement system.

Note that most other AC transient methods, such as the hot disk method [4], the laser flash method [5], and Ångström's method [6], usually respond most fundamentally to the sample's thermal diffusivity rather than conductivity. For such a diffusivity measurement, the volumetric heat capacity is also required to obtain k, which requires additional information and introduces additional uncertainty. Interestingly, because the 3ω method is based on cylindrical rather than rectilinear or spherical heat flow, its underlying thermal model directly gives the thermal conductivity rather than diffusivity.

2.2.2. *Anisotropic solids: Generalizations of the 3ω method*

(A) *Sample surface aligned with the material's principal axes*

Borca-Tasciuc and Chen extended the 3ω method to solids with anisotropic thermal conductivity [7], as long as the sample's principal axes are parallel and perpendicular to the surface of the solid (Fig. 2.4(a)). Under the same basic requirements as for an isotropic substrate listed above, i.e., a long narrow heater line periodically heating a semi-infinite substrate, Eqs. (2.2) and (2.4) can be generalized for an aligned anisotropic substrate. It turns out the only change is to replace the isotropic k with the geometric mean of the two principal thermal conductivities, k_{I} and k_{II}, as follows:

$$\sqrt{k_{\mathrm{I}} \cdot k_{\mathrm{II}}} = \frac{1}{4\pi l} \frac{I_{1\omega,\mathrm{rms}}^3 \, R_{e,0}}{[dV_{3\omega,\mathrm{rms},x}/d(\ln f)]} \frac{dR_e}{dT}, \tag{2.5}$$

$$\sqrt{k_{\mathrm{I}} \cdot k_{\mathrm{II}}} = \frac{1}{8l} \frac{I_{1\omega,\mathrm{rms}}^3 R_{e,0}}{V_{3\omega,\mathrm{rms},y}} \frac{dR_e}{dT}. \tag{2.6}$$

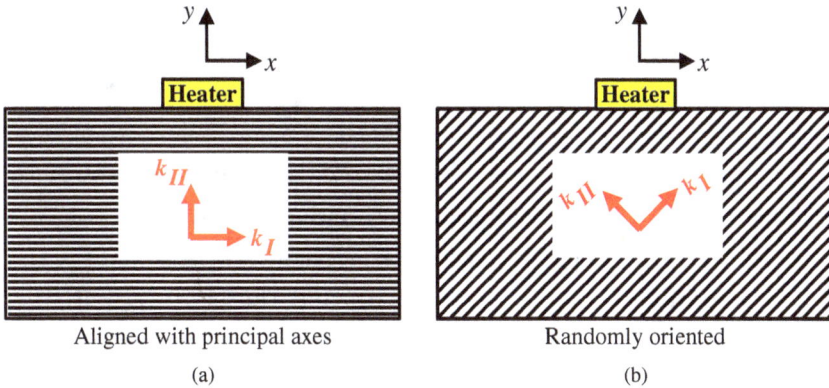

Aligned with principal axes
(a)

Randomly oriented
(b)

Fig. 2.4 The 3ω method can be extended to anisotropic substrates, to determine the values of the thermal conductivity tensor for a sample whose principal axes are (a) aligned with the surface of the sample [7], and (b) not aligned [8]. Here the heavy black lines represent the orientation of the sample's crystal structure, for example, the basal planes of graphite. See text.

To isolate k_I and k_II from their products in Eq. (2.5), the magnitude of the in-phase temperature oscillations (see [7, Eq. (10)]) versus $\ln(f)$ has to be taken into account as well.

(B) *Sample surface in an arbitrary orientation*

Recently, Mishra and Dames [8] removed the requirement for the material's principal axes to be parallel and perpendicular to the surface of the solid, thereby adapting the 3ω method to anisotropic solids with arbitrary orientations (Fig. 2.4(b)). Equations (2.5) and (2.6) are further generalized to

$$\sqrt{k_{xx} \cdot k_{yy} - k_{xy}^2} = \frac{1}{4\pi l} \frac{I_{1\omega,\mathrm{rms}}^3 R_{e,0}}{[dV_{3\omega,\mathrm{rms},x}/d(\ln f)]} \frac{dR_e}{dT}, \tag{2.7}$$

$$\sqrt{k_{xx} \cdot k_{yy} - k_{xy}^2} = \frac{1}{8l} \frac{I_{1\omega,\mathrm{rms}}^3 R_{e,0}}{V_{3\omega,\mathrm{rms},y}} \frac{dR_e}{dT}, \tag{2.8}$$

where the x and y axes are naturally defined by the surface of the solid (not to be confused with x and y indicating the in- and out-of-phase components of a periodic signal), and the k_{ij} are components of the thermal conductivity tensor in the x–y coordinate system

$$\begin{bmatrix} k_{xx} & k_{xy} \\ k_{xy} & k_{yy} \end{bmatrix}. \tag{2.9}$$

In general this tensor is not aligned with the principal axes of the crystal structure which has a diagonal conductivity tensor $\begin{bmatrix} k_\mathrm{I} & 0 \\ 0 & k_\mathrm{II} \end{bmatrix}$. It is important in Eqs. (2.7) and (2.8) to recognize that $k_{xx} \cdot k_{yy} - k_{xy}^2$ is the determinant of the tensor (2.9), and thus is invariant upon rotation of the sample in the x–y plane. Therefore, more information is needed to pin down all three components of the tensor (2.9). This can be achieved using another one or two heater lines with different orientations and taking into account the magnitude of the temperature oscillations as well as their slope versus $\ln(f)$ (see [8, Figs. 5–7]).

2.3. Thin films

The thermal properties of thin films have been intensely studied over the past three decades, and the technical details of thin-film 3ω measurements are well reviewed in the literature [2, 9–11]. In this section, we present various electrothermal techniques for different orientations. For cross-plane measurements we focus on a 3ω method, while for in-plane measurements we present six different options.

2.3.1. *Cross-plane: 3ω methods*

For a low-k film on a high-k substrate, the film's cross-plane k is relatively easy to measure, since the film is mechanically supported on a substrate which makes the microfabrication easier than a suspended configuration. The classic 3ω method for bulk materials was extended to measure thin films by Cahill *et al.* [12].

Consider the film-on-substrate schematic in Fig. 2.5. Assuming one-dimensional (1D) heat conduction vertically through the film,

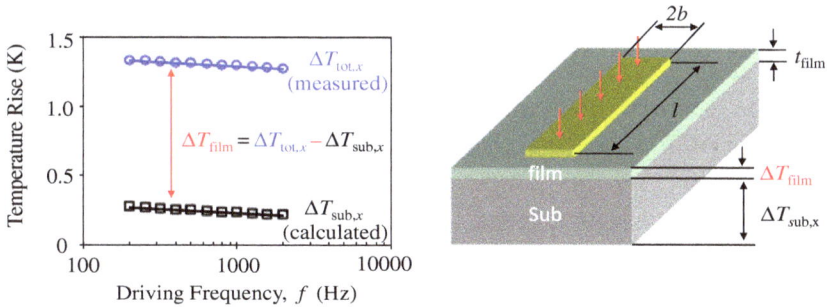

Fig. 2.5 A typical 3ω measurement performed on a 300-nm thick amorphous SiO_2 thin film on top of a doped Si wafer at $310\,K$. The points are experiments (blue) and calculations (black), respectively, while the straight lines are used to guide the eye. The temperature drop across the SiO_2 film ($\Delta T_{film} = \Delta T_{tot,x} - \Delta T_{sub,x}$) is determined by directly measuring the temperature of the heater line ($\Delta T_{tot,x}$) using Eq. (2.12), and calculating the temperature response of the substrate ($\Delta T_{sub,x}$) using the thermal model (2.13). Both $\Delta T_{tot,x}$ and $\Delta T_{sub,x}$ are for the in-phase component (subscript x) of the 3ω voltages.

we have

$$k_{\text{film},\perp} = \frac{Q t_{\text{film}}}{\Delta T_{\text{film}} 2bl},$$ (2.10)

where t_{film} is the thickness of the thin film, b and l as usual are the half-width and length of the heater line, Q is the Joule heating power applied to the heater, $\perp$ means the cross-plane direction, and ΔT_{film} is the temperature difference across the thin film.

Equation (2.10) approximates the film as being in its low-frequency, quasi-static limit, which is well satisfied for $L_{p,\text{film}} > 2.5 t_{\text{film}}$ [2, 13]. Further, to justify the assumption of 1D heat conduction in the thin film requires $b \gg t_{\text{film}}$ (by at least a factor of ~5; see [2, Table 3]), which is an additional condition beyond $b \ll L_{p,\text{sub}} \ll \min(t_{\text{sub}}, l)$ given in Sec. 2.2.1.

The key to evaluating Eq. (2.10) for $k_{\text{film},\perp}$ is to determine ΔT_{film}, and several approaches have been taken. The simplest case is a moderately-thick film ($t_{\text{film}} \sim 100\,\text{nm}$ or more) of low thermal conductivity on a large substrate ($t_{\text{sub}} \sim 500\,\mu\text{m}$ or more) of high thermal conductivity, measured in the preferred regime $t_{\text{film}} \ll b \ll L_{p,\text{sub}} \ll \min(t_{\text{sub}}, l)$. Here the thermal transfer function of the system (the temperature response of the heater line per unit heater power) can be modeled as the thermal resistance of the film, a real number, in series with the thermal impedance of the substrate, a complex number. Thus, the in-phase signal ($V_{3\omega,\text{rms},x}$) is a response to both the substrate and the thin film, while the out-of-phase signal ($V_{3\omega,\text{rms},y}$) is only a response to the substrate (see Fig. 2.3 and the corresponding discussion under Eq. (2) in [12]).

In the simple scenario shown in Fig. 2.5, in which there is only a single thin film on top of a thick substrate, we have

$$\Delta T_{\text{film}} = \Delta T_{\text{tot},x} - \Delta T_{\text{sub},x}.$$ (2.11)

Here $\Delta T_{\text{tot},x}$ is the *amplitude* of the total temperature difference across the whole system, from the heater line to some far-field T_∞, which can be determined from the first harmonic voltage, $V_{1\omega,\text{rms},x}$, and the in-phase component of the third harmonic voltage,

$V_{3\omega,\text{rms},x}$, as [1, 14]

$$\Delta T_{\text{tot},x} = -\frac{2}{\alpha} \frac{V_{3\omega,\text{rms},x}}{V_{1\omega,\text{rms},x}}. \tag{2.12}$$

The minus sign in Eq. (2.12) arises because there is a 180° phase difference between the 1ω and 3ω signals, such that the lock-in amplifier will report $V_{3\omega,\text{rms},x}$ as a negative number.

It is generally not considered practical to directly measure $\Delta T_{\text{sub},x}$, although in principle this could be achieved by incorporating a second metallic thermometer line aligned directly beneath the upper heater line. Instead, in this measurement regime a common practice is to *calculate* the magnitude of the temperature difference across the substrate alone as

$$\Delta T_{\text{sub},x} = \frac{Q}{\pi l k_{\text{sub}}} \left[\ln \left(\frac{L_p}{b} \right) + \eta \right], \tag{2.13}$$

where b and l are the half-width and length of the heater line, Q is the Joule heating power applied to the heater, $\eta \approx 0.923$ is a constant [15, 16], $L_p = \sqrt{D_{\text{sub}}/\omega_H}$ is the penetration depth defined in Eq. (2.1), and k_{sub} is the thermal conductivity of the substrate which itself can be determined by analyzing the measured $V_{3\omega,\text{rms},x}$ data with the slope method (Sec. 2.2.1). Once k_{sub} is known, the thermal diffusivity can be calculated as $D_{\text{sub}} = k_{\text{sub}}/C_{\text{sub}}$, where C_{sub} of a film is generally well approximated by its bulk handbook value, at least for characteristic lengths down to $\sim 10\,\text{nm}$ (see Appendix G).

This solution for bulk samples (Eq. (2.13)) reminds us of the thermal resistance of a cylindrical shell with inner and outer radii r_i and r_o [17],

$$R_{\text{cylind}} = \frac{1}{2\pi l k_{\text{sub}}} \ln \left(\frac{r_o}{r_i} \right), \tag{2.14}$$

where the 2 in the denominator is for a full cylinder, and would become a 1 for a hemi-cylinder. The comparison between Eqs. (2.13) and (2.14) suggests that we may visualize the classic 3ω problem as a modified cylindrical heating problem with $r_i \approx b$ and $r_o \approx L_p$, the latter of which is tunable by the heating frequency [2].

Example 2.1. Beginning with the results from Example 1.1 for the magnitudes of the 1ω and 3ω voltages, derive Eq. (2.12).

Solution: Recall the 1ω and 3ω components of voltage from Example 1.1:

$$V_{1\omega} = \left[I_{1\omega} R_{e,0} + \frac{3}{4} \alpha I_{1\omega}^3 R_{e,0}^2 R \right] \sin(\omega t),$$

$$V_{3\omega} = - \left[\frac{1}{4} \alpha I_{1\omega}^3 R_{e,0}^2 R \right] \sin(3\omega t),$$

where $I_{1\omega}$ is the amplitude of the current. Converting everything to in-phase rms values, which are conveniently recorded by lock-in amplifiers:

$$\sqrt{2} V_{1\omega,\mathrm{rms},x} = \left[\sqrt{2} I_{1\omega,\mathrm{rms},x} R_{e,0} + \frac{3}{4} \alpha (\sqrt{2} I_{1\omega,\mathrm{rms},x})^3 R_{e,0}^2 R \right],$$

$$\sqrt{2} V_{3\omega,\mathrm{rms},x} = - \left[\frac{1}{4} \alpha (\sqrt{2} I_{1\omega,\mathrm{rms},x})^3 R_{e,0}^2 R \right].$$

Note here the driving current does not have an out-of-phase component, and thus

$$I_{1\omega,\mathrm{rms},x} = I_{1\omega,\mathrm{rms}}.$$

Cleaning up,

$$V_{1\omega,\mathrm{rms},x} = \left[I_{1\omega,\mathrm{rms}} R_{e,0} + \frac{3}{2} \alpha I_{1\omega,\mathrm{rms}}^3 R_{e,0}^2 R \right],$$

$$V_{3\omega,\mathrm{rms},x} = - \left[\frac{1}{2} \alpha I_{1\omega,\mathrm{rms}}^3 R_{e,0}^2 R \right].$$

Finally we arrive at

$$\frac{V_{1\omega,\mathrm{rms},x}}{V_{3\omega,\mathrm{rms},x}} = - \frac{2}{\alpha I_{1\omega,\mathrm{rms}}^2 R_{e,0} R} - 3$$

$$= - \frac{2}{\alpha \Delta T_{\mathrm{tot},x}} - 3.$$

Since the temperature coefficient α for most metals is on the order of $10^{-3} \, \mathrm{K}^{-1}$ and temperature oscillations 1–10 K, we neglect the second

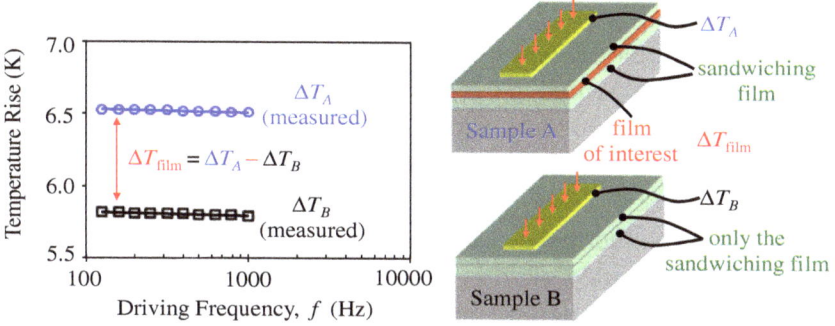

Fig. 2.6 A typical differential 3ω measurement performed on two stacks, which are nominally identical except for the film of interest incorporated within Sample A. The points are experiments, while the straight lines are used to guide the eye. Here the basic stack in Sample B includes a 28 nm-thick top SiO_2 layer, a 300 nm-thick bottom SiO_2 layer, and a doped Si substrate. The film of interest in Sample A is a 3 nm-thick graphene layer [18], sandwiched between SiO_2 layers of the same 28 nm and 300 nm thicknesses as stack B. ΔT_A and ΔT_B are related to the measured 3ω voltages using Eq. (2.12). Note that for this sample, $\Delta T_{film} = \Delta T_A - \Delta T_B$ is dominated by the thermal contact resistances between graphene and top/bottom oxide layers, instead of the thermal resistance of the graphene layer itself. Measurement temperature $T_\infty = 310$ K.

term on the right-hand side, and obtain

$$\frac{V_{1\omega,\text{rms},x}}{V_{3\omega,\text{rms},x}} \simeq -\frac{2}{\alpha \Delta T_{\text{tot},x}},$$

which is Eq. (2.12).

Another common measurement scenario is shown in Fig. 2.6, in which the film of interest is sandwiched within a more complicated stack. Here a differential 3ω method [7, 18] is recommended to extract the thermal resistance of the film of interest. In this case, we prepare two samples, A and B, with nominally identical configurations and heater patterns, except that sample A includes the thin film of interest while sample B does not. Applying the same Joule heating to both, we can extract the temperature difference across the thin film of interest as

$$\Delta T_{\text{film}} = \Delta T_A - \Delta T_B, \tag{2.15}$$

where ΔT_A and ΔT_B are the amplitudes of the temperature oscillations of the heater lines in the two samples, both of which can be obtained using Eq. (2.12) by measuring the first and third harmonic voltages of samples A and B.

The results presented in Fig. 2.6 also highlight two other aspects of the differential 3ω method. First, the differential thermal resistance contributed by incorporating a new thin film into a stack will in general involve two contributions, the conduction resistance through the film and the thermal contact resistance between the new film and the neighboring sandwich layers. The thinner the film of interest, the more the latter will dominate, which was the case in [18]. Second, to better ensure 1D cross-plane heat flow through the film stack, the stack can be microfabricated as a mesa of width $2b$ aligned carefully to the heater line, which was accomplished in [18] by a self-aligned ion milling step, as depicted in Fig. 2.7 and discussed below.

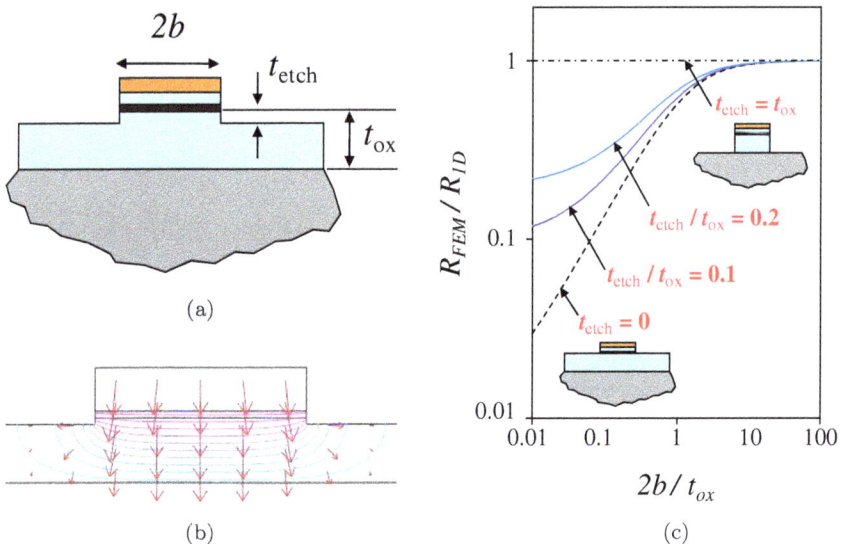

Fig. 2.7 Justification of one-dimensional (1D) heat transfer. (a) Three key length scales of the problem. (b) Visualization of the isotherms and flux lines using a 2D FEM simulation, approximating the substrate as an isothermal boundary condition. (c) Convergence of the actual thermal resistance to the ideal 1D resistance, $R_{\text{th,FEM}}/R_{\text{th,1D}}$, as a function of the dimensionless group $2b/t_{\text{ox}}$. Typical expected values of these parameters in the real experiments: $2b = 3\,\mu\text{m}$, $t_{\text{etch}} = 60\,\text{nm}$, $t_{\text{ox}} = 300\,\text{nm}$.

As an example of ensuring 1D heat transfer through a film stack, we analyze the sandwich structure (top SiO_2 + graphene + bottom SiO_2) in [18]. As indicated in Fig. 2.7(a), there are three important length scales in this problem: the heater width ($2b$), the thickness of the lower oxide (t_{ox}), and the etching depth (t_{etch}). Ideally, to ensure 1D heat transfer through the sandwich structure we would like $2b \gg t_{ox}$ or $t_{etch} = t_{ox}$. In practice, however, we are constrained by the graphene flake size and concerned about the time and cost of the ion milling. To provide a quantitative basis for the inevitable compromise between ideality and reality, we simulated various nonideal structures using a two-dimensional (2D) finite element method (COMSOL FEM). To mimic the real experiment, we use a flux boundary condition to represent the heater. We then apply a temperature boundary condition on the lower surface of the bottom oxide to represent the transition to the high-k substrate, and set the other boundaries to be adiabatic.

Figure 2.7(b) shows the isotherms and adiabats for a representative structure, which confirms the 1D heat transfer qualitatively. Note that we exaggerated the thickness of graphene for clarity. To quantify these nonideality results, in Fig. 2.7(c) we normalize the real thermal resistance (R_{FEM}) by the ideal 1D resistance, $R_{1D} = t_{ox}/(k_{ox} \cdot 2b)$, and plot it as a function of a dimensionless group $2b/t_{ox}$. We show curves for four different etch depths, from zero to complete etching through the bottom oxide. The first feature of Fig. 2.7(c) is that R_{FEM} gradually converges to R_{1D} as $2b/t_{ox}$ increases, as expected. The second feature is that as the etching depth increases, the convergence becomes faster. Based on this analysis and considering the experimental practicalities, we fixed the two adjustable microfabrication parameters to be $2b = 3\,\mu m$ and $t_{etch} = 0.2 \cdot t_{ox}$, which according to Fig. 2.7(c) introduces only a modest 4.5% error as compared to assuming purely 1D heat transfer through the sandwich structure.

2.3.2. *In-plane*

As compared to a cross-plane measurement, measurements of a film's in-plane thermal conductivity are more challenging because of the potential for very large parasitic heat leakage from the film of interest into the substrate and/or the adjacent layers. An obvious strategy to avoid this problem is to etch out the substrate underneath the film, though at the cost of much more challenging microfabrication. An alternative approach is to keep the substrate and thus live with the parasitic heat leakage, shifting the challenge to the measurement sensitivity and heat transfer analysis. In this section, we discuss six techniques, four suspended and two supported, based on these two philosophies.

(A) *Suspended microfabricated-device method*

This method achieves excellent simplicity in its model for heat transfer through the sample, but suffers from the complexity of microfabrication (recall Fig. 1.1). It was initially developed for nanotubes and nanowires [19–21], and later applied to thin films (e.g., graphene [22]). Another merit of this technique is that it does not assume diffusive transport through the suspended sample, which makes this method applicable to study ballistic thermal transport.

　　To illustrate the measurement concept, here we analyze the schematic of Fig. 2.8(a) using a simplified thermal circuit, as shown in Fig. 2.8(b). In this thermal circuit, we lumped the Joule heating of the serpentine heater (Q_{Htr}) inside the heating island with that of the two long current leads ($2Q_{\mathrm{Lead}}$) outside the island. This crudely approximates all of the Joule heating in each of the two current leads $Q_{\mathrm{Lead}} = I^2 R_{e,\mathrm{Lead}}$, where $R_{e,\mathrm{Lead}}$ is the electrical resistance of a single metal lead, as if it were localized at the heating island, an approximation which is revisited briefly below.

　　We first look at the thermal path connecting the heating island to the ambient through the five long beams, and obtain

$$Q_1 \cdot R_{\mathrm{Beam}} = T_{\mathrm{Htr}} - T_\infty, \qquad (2.16)$$

where R_{Beam} is the total (parallel) thermal resistance of the five long beams supporting the heating island, Q_1 is the corresponding heat flowing through this path, and T_{Htr} and T_∞ are the temperatures of the heating island and ambient, respectively.

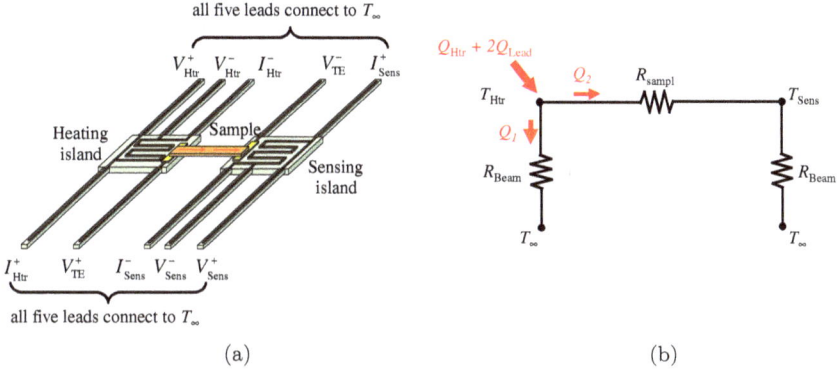

Fig. 2.8 (a) Schematic of the suspended microfabricated-device method, repeated from Fig. 1.3 for convenience. (b) Corresponding simplified thermal circuit. The node marked T_{Htr} experiences the Joule heating of the island's serpentine heater plus that of the two long current leads. The latter term is an approximation which is revisited and improved in Eq. (2.20). R_{Beam} represents the parallel combination of five individual supporting legs.

We next analyze the thermal path from the heater island to the sensing island, and sensing island to the thermal ground, and obtain

$$Q_2 \cdot R_{\text{sampl}} = T_{\text{Htr}} - T_{\text{Sens}},$$
$$Q_2 \cdot R_{\text{Beam}} = T_{\text{Sens}} - T_\infty, \tag{2.17}$$

where R_{sampl} is the thermal resistance of the sample, Q_2 is the corresponding heat flowing through this path, and T_{Sens} is the temperature of the sensing island. Note here we assume R_{Beam} is the same for both heating and sensing islands, because of the identical design for the microfabrication.

Finally, by energy conservation, we have

$$Q_1 + Q_2 = Q_{\text{Htr}} + 2Q_{\text{Lead}}. \tag{2.18}$$

Combining the above equations, we arrive at

$$R_{\text{sampl}} = \left[\frac{(T_{\text{Htr}} - T_\infty) + (T_{\text{Sens}} - T_\infty)}{Q_{\text{Htr}} + 2Q_{\text{Lead}}} \right] \left[\frac{T_{\text{Htr}} - T_{\text{Sens}}}{T_{\text{Sens}} - T_\infty} \right]. \tag{2.19}$$

Here everything on the right-hand side is known: the temperatures can be measured by the four-probe resistance thermometry and the

Joule heating can be calculated by $Q_{\text{Htr}} = I^2 R_{e,\text{Serpentine}}$ and $Q_{\text{Lead}} = I^2 R_{e,\text{Lead}}$. Note also that R_{Beam} has canceled out.

An improved analysis [21, 23] takes into account the fact that the Joule heating in the current leads of the heating island is distributed instead of lumped. This updates Eq. (2.19) to

$$R_{\text{sampl}} = \left[\frac{(T_{\text{Htr}} - T_\infty) + (T_{\text{Sens}} - T_\infty)}{Q_{\text{Htr}} + Q_{\text{Lead}}} \right] \left[\frac{T_{\text{Htr}} - T_{\text{Sens}}}{T_{\text{Sens}} - T_\infty} \right], \quad (2.20)$$

which is the recommended form.

The derivation above assumes that the radiation and convection losses from the suspended samples are negligible, which, however, is not always ensured. Although for a measurement conducted in a cryostat with vacuum level better than 10^{-5} Torr (see Appendix E), the convection loss can be safely neglected, the radiation loss may require some care (see Sec. 3.2.1).

Another key assumption of this technique is the spatial uniformity in temperatures of the heating and sensing islands. This is ensured by thermal design for a small generalized Biot number, which requires

$$R_{\text{island}} \ll \min(R_{\text{sampl}}, R_{\text{Beam}}), \quad (2.21)$$

where R_{island} is the internal thermal resistance of the heating and sensing islands.

There are several practical challenges in implementing this technique. First, it requires some method to place a sample between the heating and sensing islands. Common tricks include micromanipulation of nanostructures [23, 24], drop-casting from a solution containing suspended nanostructures [23], *in-situ* growth of nanostructures on fabricated microdevices [25], and integrated microfabrication of microdevices and nanostructures [26]. Second, the thermal contact between the sample and the two islands also requires extra attention, and there are several methods to address this issue [19, 21, 23, 27].

(B) *Distributed self-heating method*

This technique requires the sample to be electrically conducting with good ohmic contacts and a stable temperature coefficient of resistance. It also requires suspending the thin film or nanostructure, although the microfabrication may be somewhat simpler than the other suspended techniques. As can be seen from Example 1.2 and below, the heat transfer model of this technique relies on Fourier's law, thus making it inapplicable to study ballistic thermal transport.

For a uniformly distributed volumetric heat source in a long filamentary sample, recall the average temperature across the sample obtained in Example 1.2:

$$\langle T \rangle = T_\infty + \frac{IVL}{12k_{\text{film},\|}A_{\text{cr}}}, \tag{2.22}$$

which corresponds to an equivalent thermal resistance of the thin film,

$$R_{\text{film}} = \frac{L}{12k_{\text{film},\|}A_{\text{cr}}}, \tag{2.23}$$

where k, L, and A_{cr} are the thermal conductivity, length, and cross-sectional area of the thin film and $Q = IV$ is the Joule heating power. Note the factor of $1/12$, which arises because of the distributed heating and temperature averaging.

Thus, the thermal conductivity of the thin film is

$$k_{\text{film},\|} = \frac{IVL}{12(\langle T \rangle - T_\infty)A_{\text{cr}}}. \tag{2.24}$$

As shown schematically in Fig. 2.9(b), after measuring the I–V curves of the suspended thin film, we plot the electrical resistance

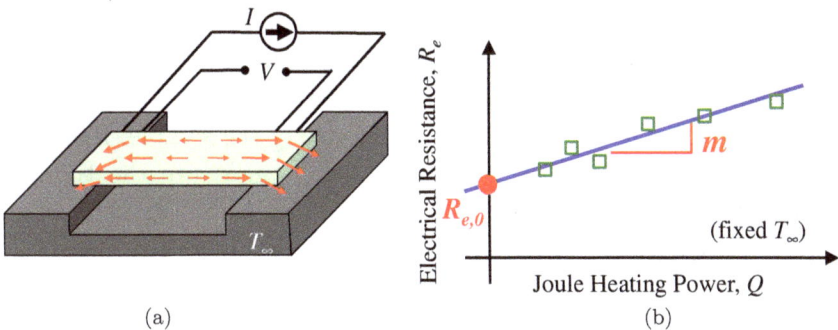

Fig. 2.9 (a) Schematic of the distributed self-heating method, repeated from Fig. 1.4 for convenience. (b) Illustration of the data processing. The thermal conductivity of a suspended thin film or nanotube/nanowire can be extracted from the slope m of a plot (schematic points) of electrical resistance vs. heating power (Eq. (2.25)).

$(R_e = V/I)$ of the thin film as a function of the Joule heating power. Importantly, these I–V curves are deliberately driven into a regime where the self-heating is substantial, in contrast to standard practice for resistance thermometry (e.g., as depicted below in Fig. 2.14(b)).

Finally, the thermal conductivity of the thin film can be extracted as

$$k_{\text{film},\parallel} = \frac{\alpha R_{e,0} L}{12 m A_{\text{cr}}}, \qquad (2.25)$$

where m and $R_{e,0}$ are the slope and intercept of the $R_e - Q$ plot, and α is the temperature coefficient of the thin film.

Several important aspects including the radiation losses, the contact and substrate spreading resistance, and the placement of voltage probes, require careful thermal design and are discussed elsewhere [2].

(C) *T-bridge method*

Some filamentary samples cannot be directly measured using the distributed self-heating method because they lack the necessary electrical properties identified above. In such cases, a related measurement bridges the sample across the midpoint of a separate sensing wire made of a suitable metal, forming a "T" shape (Fig. 2.10(a)) [28, 29]. As long as the thermal resistances of the sample (R_{sampl}) and sensing wire ($R_{\text{Htr}} = L/(k_{\text{Htr}} A_{\text{cr}})$) are comparable (within a factor of $\sim$10; see below), the spatially-averaged temperature rise of the sensing wire is sensitive to the thermal conductance of the sample. Recently, this T-bridge method has been extended to measure the thermal conductivity of 2D materials [30, 31].

To analyze this quantitatively, recall the modified parabolic temperature profile obtained in Example 1.3,

$$T(x) = T_\infty + \frac{Q}{8} R_{\text{Htr}} \left[-\left(\frac{x}{L/2}\right)^2 + \frac{\gamma}{1+\gamma}\left|\frac{x}{L/2}\right| + \frac{1}{1+\gamma} \right],$$

$$(2.26)$$

where $\gamma = R_{\text{Htr}}/(4 R_{\text{sampl}})$.

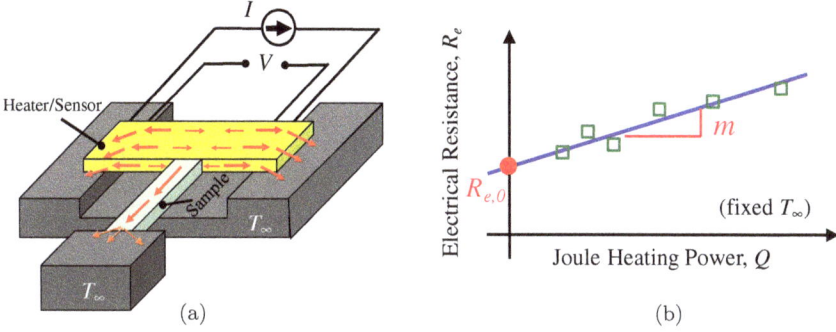

Fig. 2.10 (a) Schematic of the T-bridge method, repeated from Fig. 1.5 for convenience. (b) Illustration of the data processing, which is very similar to that of Fig. 2.9. Here the thermal resistance of the distributed sensing wire (gold structure in (a)) must be known, and slope of the $R - Q$ plot can be related to the thermal resistance of the suspended thin film or nanotube/nanowire (green structure in (a)) using Eq. (2.29).

Averaging temperature along the sensing wire gives

$$\langle T \rangle = T_\infty + \frac{1}{48} Q R_{\text{Htr}} \left(1 + \frac{3}{\gamma + 1} \right). \tag{2.27}$$

Using the relationship between the electrical resistance of the sensing wire and the temperature, $R_e = R_{e,0} + \alpha R_{e,0} \Delta T$, with $\Delta T = \langle T \rangle - T_\infty$, we can extract the key parameter γ from the slope of the $R_e - Q$ plot (Fig. 2.10(b)):

$$\gamma = \frac{4 \alpha R_{e,0} R_{\text{Htr}} - 48m}{48m - \alpha R_{e,0} R_{\text{Htr}}}, \tag{2.28}$$

where m and $R_{e,0}$ are the slope and intercept of the $R_e - Q$ plot. Thus, the thermal resistance of the sample is

$$R_{\text{sampl}} = \frac{R_{\text{Htr}}}{4} \frac{48m - \alpha R_{e,0} R_{\text{Htr}}}{4 \alpha R_{e,0} R_{\text{Htr}} - 48m}. \tag{2.29}$$

It can be shown [29] that the sensitivity (defined later in Eq. (4.4)) of $\langle T \rangle$ to changes in R_{sampl} is

$$S = \frac{3}{4} (\gamma^{1/2} + \gamma^{-1/2})^{-2}, \tag{2.30}$$

which peaks at $\gamma = 1$ and falls off for both large and small γ, with tolerable sensitivity only in the range $0.1 < \gamma < 10$.

The derivation presented here assumes negligible surface heat losses by radiation and convection, the sample is perfectly centered at the midpoint of the sensing wire, and the sample diameter or width is small compared to the sensing wire length. The limits of these assumptions have been addressed in the primary literature [29, 31].

(D) *Central-line heater method*

The central-line heater method is depicted in Fig. 2.11. This scheme has one of the simplest heat transfer models, but correspondingly imposes high demands on microfabrication to suspend the thin film and create a heater line at the center of the suspended film.

As shown in Fig. 2.11, assuming perfect symmetry we can obtain the thermal resistance of the half of the thin film between the heater and one of the banks of the trench:

$$R_{\text{half-film}} = \frac{T_{\text{Htr}} - T_{\text{bank}}}{\frac{1}{2}Q_{\text{Htr}}}, \tag{2.31}$$

where Q_{Htr} is the Joule heating applied through the central heater line, and the prefactor $\frac{1}{2}$ accounts for the fact that half of the Joule heating flows to each side of the heater. T_{Htr} and T_{bank} are the temperatures of the central heater line and the sensor on the bank of the trench, respectively, measured using four-probe resistance thermometry.

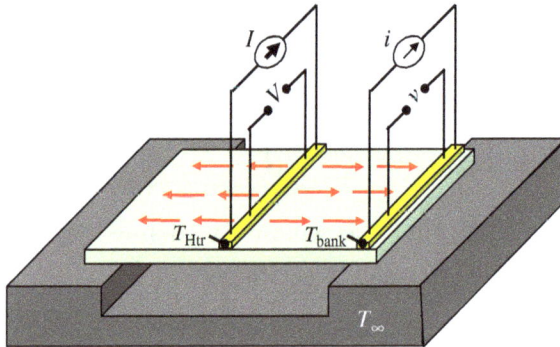

Fig. 2.11 Schematic of the central-line heater method, repeated from Fig. 1.6 for convenience. The second metal line (i, v) is strictly for thermometry, not heating.

Note that Eq. (2.31) does not make any assumption about the thermal transport, diffusive vs. ballistic, so it applies to both regimes. Assuming diffusive transport, knowing the geometry of the film we can extract the thermal conductivity

$$k_{\text{film},\|} = \frac{\frac{1}{2}Q_{\text{Htr}}}{T_{\text{Htr}} - T_{\text{bank}}} \frac{\frac{1}{2}L}{A_{\text{cr}}}, \tag{2.32}$$

where L is the length of the sample spanning the trench from bank to bank and A_{cr} is the cross-sectional area of the film.

As in the distributed self-heating method, this central-line heater method requires attention to issues such as radiation losses, thermal contact and substrate spreading resistances, 2D spreading effects, and placement of voltage probes. See [2, Sec. 5].

(E) *Variable-linewidth 3ω method*

In Sec. 2.3.1, we saw how to use a 3ω method to measure the cross-plane thermal conductivity of a film by assuming 1D heat conduction across the thin film. Here we relax this assumption, and thereby extend the 3ω method to measure the in-plane thermal conductivity of thin films as well [7].

The approximation of 1D heat conduction through the film overestimates the cross-plane thermal conductivity, since in actual experiments it mistakenly treats the lateral heat spreading near the heater edges as vertical heat flow, corresponding to an erroneously large $k_\perp$. This edge effect exists even for a wide heater on top of a thin film (Fig. 2.12(a)), but is negligible for $b \gg t$.

Solving the 2D heat diffusion equation for more general material systems allowing anisotropic properties of both thin film and substrate (aligned with the principal axes), Borca and Chen were able to take into account this lateral heat spreading effect [7]. For the typical scenario in which the thermal conductivity of the substrate is much larger than that of the thin film, they found a convenient correction factor:

$$\frac{k_{\text{film},\perp}}{k_{1D}} = \frac{1}{1 + 0.38\xi}, \tag{2.33}$$

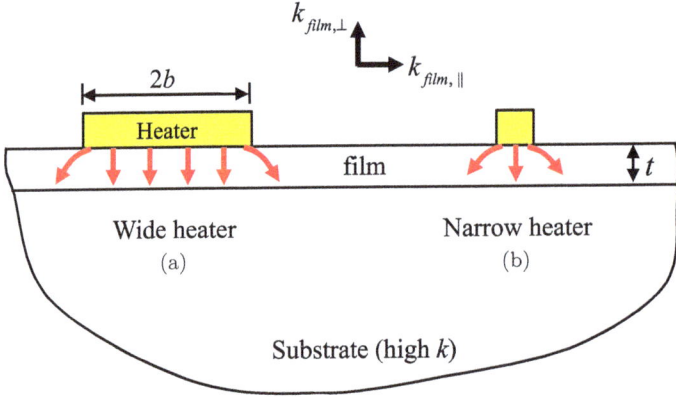

Fig. 2.12   Concept of the variable-linewidth 3ω for thin film measurement. (a) When the heater linewidth is much larger than the film thickness ($b \gg t$), the heat transfer is most sensitive to the film's cross-plane thermal conductivity ($k_{\text{film},\perp}$). (b) When $b < t$, the heat transfer becomes sensitive to the in-plane thermal conductivity ($k_{\text{film},\parallel}$) as well.

where $k_{\text{film},\perp}$ is the actual cross-plane thermal conductivity of the film, k_{1D} is the apparent measured cross-plane thermal conductivity under the assumption of 1D heat conduction across the thin film, and $\xi = \sqrt{k_{\text{film},\parallel}/k_{\text{film},\perp}}(t/b)$ can be viewed as the (square root of the) ratio between the in-plane and cross-plane thermal conductance. Equation (2.33) requires $\xi < 10$ for error of 3% or less, which is almost always the case though this also implies limited sensitivity to $k_{\text{film},\parallel}$. A more general expression for any ξ is derived in [7], which involves numerical evaluation of an integral.

Equation (2.33) offers a way to measure both $k_{\text{film},\perp}$ and $k_{\text{film},\parallel}$ using the 3ω method. First, as shown in Fig. 2.12(a), a wide heater with $\xi \ll 1$ is used to extract $k_\perp$, since in this scenario the 1D approximation is well justified. Next, a narrow heater (Fig. 2.12(b)) with larger ξ is used to extract $k_{\text{film},\parallel}$ using Eq. (2.33), using the $k_{\text{film},\perp}$ obtained earlier. Because this approach is always less sensitive to $k_{\text{film},\parallel}$ than $k_{\text{film},\perp}$, we recommend that more than two different heater widths should be included in the study, the narrowest of which should reach $\xi \geq 3$.

(F) *Heat spreader method*

The heat spreader method is another technique that relaxes the constraints on microfabrication but adds complexity in heat transfer analysis.

Recall Sec. 1.7 and the schematic of the heat spreader method shown in Fig. 1.8. The in-plane thermal conductivity of the sandwiched thin film can be obtained by fitting measured experimental quantities (the joule heating of the heater and temperature of the three sensors) to a suitable heat transfer model, which also involves various other known geometries and thermal properties. Here we discuss two such models.

Recognizing that the thermal conductivity of the Si substrate is very high such that it may approximate a perfect heat sink, the simplest thermal model is to treat the thin film (graphene in Fig. 1.8) as a fin, where the effective "convection" coefficient h_{eff} represents the vertical conduction through the lower SiO$_2$ layer into the Si heat sink. The interfacial contact resistances may also be important. Thus,

$$h_{\text{eff}} = (t_{\text{BotOx}}/k_{\text{BotOx}} + R''_{c,\text{gr-ox}} + R''_{c,\text{ox-Si}})^{-1}, \qquad (2.34)$$

where k_{BotOx} and t_{BotOx} are the thermal conductivity and thickness of the lower oxide layer, and the two R''_c terms are the specific contact resistances (with SI units m$^2\cdot$K/W) from graphene-to-oxide and from oxide-to-silicon, respectively.

Using textbook fin theory (see Example 1.4), we estimate the characteristic fin length,

$$\beta^{-1} = \sqrt{k_{\|} A_c/h_{\text{eff}} P}, \qquad (2.35)$$

where $k_{\|}$ is the in-plane thermal conductivity, and A_c and P are the cross-sectional area and "wetted perimeter" of the graphene fin. For a graphene flake of width w and thickness t_{gr}, we have $A_{\text{cr}} = w \cdot t_{\text{gr}}$ and $P = w$ rather than $2w$, because the flake only conducts heat out through its lower face.

The textbook fin temperature profile is $T(x) = T_\infty + T_0 \cdot \exp(-\beta \cdot x)$. Thus, by fitting this profile to the experimental $T(x)$ profile measured using the three temperature sensors (Fig. 1.8), we can extract the fin length β^{-1}, and thus the thermal conductivity $k_\parallel$ from Eq. (2.35).

However, this textbook 1D fin treatment assumes the fin length $\beta^{-1} \gg t_{\mathrm{BotOx}}$ to justify Eq. (2.34), which might not be satisfied in real experiments. For example, in our measurements of the thermal conductivity of encased graphene [32], the 12-layer-thick graphene sample has a fin length of $\beta^{-1} \simeq 310\,\mathrm{nm}$ which is comparable to the thickness of the lower oxide layer of $t_{\mathrm{BotOx}} = 320\,\mathrm{nm}$, and thus violating the assumption $\beta^{-1} \gg t_{\mathrm{BotOx}}$. Here the parameters used for the estimate above are: $k_{\mathrm{gr},\parallel} = 92\ \mathrm{W/m \cdot K}$ [32], $k_{\mathrm{BotOx}} = 1.4\ \mathrm{W/m \cdot K}$ [32], $t_{\mathrm{gr}} = 4.1\,\mathrm{nm}$ [32], $t_{\mathrm{BotOx}} = 320\,\mathrm{nm}$ [32], $R''_{c,\mathrm{gr-ox}} = 9.0 \times 10^{-9}\ \mathrm{m^2 \cdot K/W}$ [18], and $R''_{c,\mathrm{ox-Si}} = 2.0 \times 10^{-8}\ \mathrm{m^2 \cdot K/W}$ [15].

Although in our experience this fin model does not usually lead to a highly accurate result, it nevertheless gave good insight to guide our experimental design. For example, recognizing the even smaller β^{-1} expected for thin graphene samples with fewer layers, we focused on sharpening the resolution of the e-beam lithography to reduce the center-to-center distance between the adjacent T sensors from 740 nm to 350 nm.

Due to the aforementioned shortcomings with the 1D fin model, to achieve satisfactory accuracy in [32] we found it necessary to use a three-dimensional (3D) finite element method (COMSOL) to simulate additional details of the structure. As shown in Fig. 2.13, we treated the unknown k of the graphene layer as an adjustable parameter, and solved the 3D FEM model iteratively to find the k that gives the best agreement between simulated and measured quantities (the temperature rise at the three sensors, normalized by heater power). This nonlinear least-squares fitting process was automated by using MATLAB to interface directly with COMSOL. We validated this scheme by two control experiments [32]: one experiment removing the graphene film from Fig. 2.13, and the other replacing the graphene film with a 38-nm thick Pt thin film. These two control experiments agreed with other measurements using

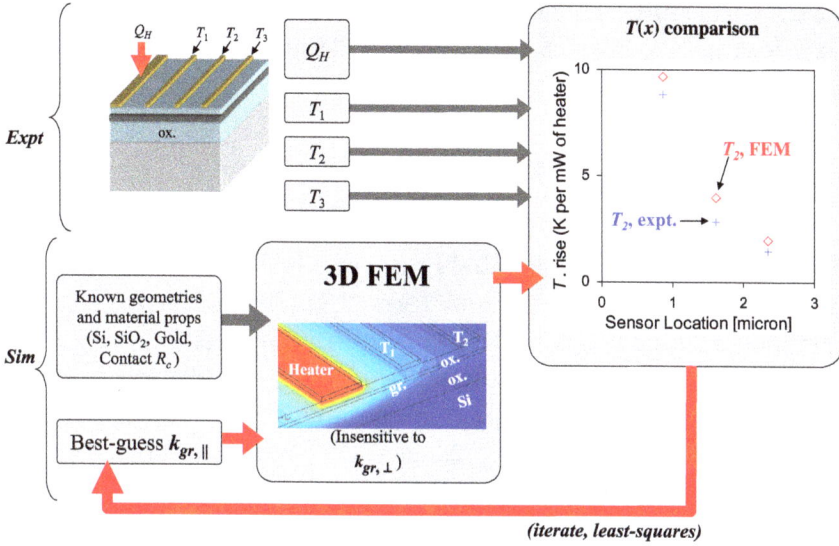

Fig. 2.13 Fitting algorithm to extract the in-plane thermal conductivity of encased graphene using the heat spreader method of Fig. 1.8 [32].

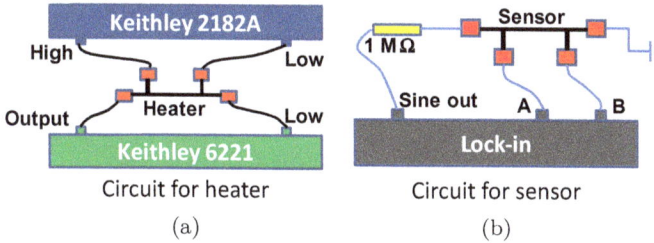

Fig. 2.14 Schematic electrical circuit to apply heating power and measure the thermal response of (a) heater and (b) temperature sensors in the heat spreader method. Here (a) is a DC method while (b) is AC.

different methods to better than 1% (3ω) and 5% (Wiedemann–Franz Law, Sec. 3.4), respectively, thus confirming the accuracy of this heat spreader + 3D FEM scheme.

The electrical connections for the heater and the temperature sensors are shown in Fig. 2.14. The heater is driven by a DC current source (Keithley 6221) and detected by a nanovoltmeter (Keithley

2182A). Each thermometer is driven and detected at AC by its own lock-in amplifier (SR830 and/or SR850), using small currents to ensure negligible self-heating.

2.4.   Suspended 1D structures: Nanotubes and nanowires

The suspended microfabricated-device method (Sec. 2.3.2(A)), the distributed self-heating method (Sec. 2.3.2(B)), and the T-bridge method (Sec. 2.3.2(C)) for the in-plane thermal conductivity measurement of thin films can also be directly applied to measure nanotubes and nanowires. For these three techniques, we simply refer the readers back to Sec. 2.3.2(A–C) for details. Note that the thermal contact between nanotubes/nanowires and the measurement platforms requires extra care, since now it is in theory a line contact, which is even worse than the plane contact in the thin film scenario.

The central-line heater method has not been applied for nanotubes/nanowires measurements for the obvious reason that the electrodes in Figs. 2.10 and 2.11 cannot be placed on top of a single 1D structure. Nor to our knowledge has the heat spreader method, though in principle it might be viable for a large diameter MWCNT with very high k.

A variation of the 3ω method (Sec. 2.2.1) can also be used to measure nanotubes and nanowires, with the heat transfer model modified to describe a suspended 1D structure. We will now discuss this in more detail.

(A) 3ω method for suspended 1D structures

This technique requires the sample to be electrically conducting, or to be coated with an electrically conducting layer if the sample itself is a poor conductor. By varying the frequency of the driving current, this technique can extract both the thermal conductivity and the heat capacity. Note that this technique can be viewed as the periodic version of the distributed self-heating method as described in Sec. 2.3.2(B).

The steady periodic solution for the temperature response of a 1D structure subject to a spatially distributed periodic heat source was presented by Lu *et al.* as a series solution [33], and subsequently as a closed-form solution by Dames *et al.* [3]. The latter also gave an approximate lumped solution motivated by the similarity of the full solution to a first-order RC system, which is surprisingly accurate and thus used here for simplicity.

These solutions can be divided into low and high frequency limits depending on the dimensionless frequency $\hat{\omega} = \omega \cdot \tau$, where ω is the angular frequency of the driving current and τ is the diffusion time across the nanotube/nanowire defined as $\tau = L^2/D$, with D as the diffusivity of the nanotube/nanowire. To be consistent with the discussions throughout this book, here L is the full length of the wire, whereas L represents the half length of the wire in [3].

In the low frequency limit ($\omega \cdot \tau \ll 1$) and using the in-phase component of the 3ω voltage ($V_{3\omega,\mathrm{rms},x}$), Ref. [3] gives the thermal conductivity of the 1D structure as

$$k = -\frac{L}{24A_{\mathrm{cr}}} \frac{I_{1\omega,\mathrm{rms}}^3 R_{e,0}}{V_{3\omega,\mathrm{rms},x}} \frac{dR_e}{dT}, \qquad (2.36)$$

where L is the length of the wire, and other quantities are as defined following Eq. (2.2) (Sec. 2.2.1). As in Eq. (2.12), the minus sign in Eq. (2.36) is due to the 180° phase difference between the 1ω and 3ω signals, and will be canceled by the negative value of $V_{3\omega,\mathrm{rms},x}$.

The heat capacity is also readily obtained. For example, in the high frequency limit ($\omega \cdot \tau \gg 1$) and using the out-of-phase 3ω voltage ($V_{3\omega,\mathrm{rms},y}$), Ref. [3] gives C as

$$C = \frac{5}{24LA_{\mathrm{cr}}} \frac{I_{1\omega,\mathrm{rms}}^3 R_{e,0}}{V_{3\omega,\mathrm{rms},y}} \frac{dR_e}{dT} \frac{1}{\omega}. \qquad (2.37)$$

2.5. Liquids, biological tissues, and other soft matter: Supported 3ω method

In the traditional 3ω method, the substrate functions as both sample and mechanical support for the heater line. This is inappropriate for liquids, biological tissues, and other soft matter, because depositing the 3ω heater line requires harsh microfabrication which is incompatible with samples which are soft, liquid, or otherwise chemically sensitive. For such delicate samples, a "supported" or "bi-directional" 3ω method is helpful, and has been applied to measure liquids [34–37] and biological samples including tissues as thin as $100\,\mu m$ [38] and potentially even single cells [39].

As shown in Fig. 2.15(a), a metallic heater line is first deposited on top of a rigid inorganic substrate of low k, e.g., a glass microscope slide, and coated with a thin dielectric isolation layer. The thermal conductivity of this substrate (k_{sub}) is calibrated using the classical 3ω method (Sec. 2.2.1). For example, in the ideal intermediate frequency limit $[b \ll L_p \ll \min(t_{\text{sub}}, l)]$, k_{sub} can be obtained from Eq. (2.2) using the slope method:

$$k_{\text{sub}} = \frac{1}{4\pi l} \frac{I_{1\omega,\text{sub}}^3 R_{e,0}}{[dV_{3\omega,\text{sub}}/d(\ln f)]} \frac{dR_e}{dT}. \tag{2.38}$$

Next, a soft biological sample is placed on top of the heater line, in intimate contact with the dielectric isolation layer (Fig. 2.15(b)). Approximating the substrate and the sample as two parallel thermal impedances (the "boundary mismatch approximation", BMA), we

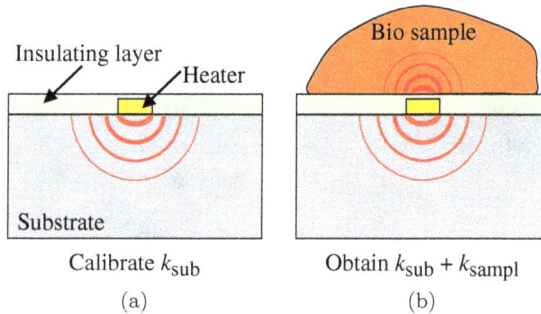

Fig. 2.15 Supported or bi-directional 3ω method to measure k of biological tissues and other delicate samples. (a) The "sensor" (substrate + heater + insulating layer) is first calibrated for k_{sub} using a classic 3ω method (Sec. 2.2.1). Harsh microfabrication processes are only used when fabricating this substrate, but not applied to the delicate sample. (b) The sample is then placed on top. Now the heat dissipation from the heater has two parallel paths, which knowing k_{sub} can be analyzed to extract k_{sampl}.

obtain

$$k_{\text{sub}} + k_{\text{sampl}} = \frac{1}{4\pi l} \frac{I^3_{1\omega,\text{sub+sampl}} R_{e,0}}{[dV_{3\omega,\text{sub+sampl}}/d(\ln f)]} \frac{dR_e}{dT}. \tag{2.39}$$

Thus, combining Eqs. (2.38) and (2.39), the thermal conductivity of the biological sample can be extracted. Note that another advantage of the supported 3ω approach is that the sensor (Fig. 2.15(a)) can be reused for multiple samples, whereas the traditional 3ω approach requires microfabrication on every sample.

The BMA corresponds to neglecting heat transfer between the sample and substrate except at the heater line itself, and is widely used [34–37] because it greatly simplifies the analysis of experimental data as compared to solving the fully coupled heat conduction problem. The errors in the BMA have been discussed in [38], who found that Eq. (2.39) becomes exact in the ideal intermediate limit $[b \ll L_p \ll \min(t_{\text{sub}}, l)]$ regardless of the diffusivity contrast between the substrate and the sample. The errors corresponding to the assumption of low frequency limit are analyzed in [38, Figs. 4 and 7].

References

[1] D. G. Cahill, "Thermal conductivity measurement from 30 to 750 K: the 3-omega method," *Rev. Sci. Instrum.* **61**(2), pp. 802–808, 1990.

[2] C. Dames, "Measuring the thermal conductivity of thin films: 3 omega and related electrothermal methods," *Annu. Rev. Heat Transf.* **16**, pp. 7–49, 2013.

[3] C. Dames and G. Chen, "1ω, 2ω, and 3ω methods for measurements of thermal properties," *Rev. Sci. Instrum.* **76**(12), p. 124902, 2005.

[4] S. E. Gustafsson, "Transient diffusivity plane source techniques for thermal conductivity and thermal diffusivity measurements of solid materials," *Rev. Sci. Instrum.* **62**(3), pp. 797–804, 1991.

[5] W. J. Parker, R. J. Jenkins, C. P. Butler and G. L. Abbott, "Flash method of determining thermal diffusivity, heat capacity, and thermal conductivity," *J. Appl. Phys.* **32**, p. 1679, 1961.

[6] C. H. Bosanquet and R. Aris, "On the application of Ångström's method of measuring thermal conductivity," *Br. J. Appl. Phys.* **5**(7), pp. 252–255, 1954.

[7] T. Borca-Tasciuc, A. R. Kumar and G. Chen, "Data reduction in 3ω method for thin-film thermal conductivity determination," *Rev. Sci. Instrum.* **72**(4), p. 2139, 2001.

[8] V. Mishra, C. L. Hardin, J. E. Garay and C. Dames, "A 3-omega method to measure an arbitrary anisotropic thermal conductivity tensor," *Rev. Sci. Instrum.* **86**(5), p. 54902, 2015.

[9] D. G. Cahill, H. E. Fischer, T. Klitsner, E. T. Swartz and R. O. Pohl, "Thermal conductivity of thin films: Measurements and understanding," *J. Vac. Sci. Technol. A* **7**, p. 1259, 1989.

[10] K. E. Goodson and M. I. Flik, "Solid layer thermal-conductivity measurement techniques," *Appl. Mech. Rev.* **47**, pp. 101–112, 1994.

[11] T. Borca-Tasciuc and G. Chen, "Experimental techniques for thin-film thermal conductivity," in *Thermal Conductivity: Theory, Properties, and Applications.* Springer, 2004.

[12] D. Cahill, M. Katiyar and J. R. Abelson, "Thermal conductivity of a-Si:H thin films," *Phys. Rev. B* **50**(9), pp. 6077–6082, 1994.

[13] Y. S. Ju and K. E. Goodson, "Process-dependent thermal transport properties of silicon-dioxide films deposited using low-pressure chemical vapor deposition," *J. Appl. Phys.* **85**(10), pp. 7130–7134, 1999.

[14] D. G. Cahill, "Erratum: Thermal conductivity measurement from 30 to 750 K: The 3-omega method," *Rev. Sci. Instrum.* **73**(10), p. 3701, 2002.

[15] S. Lee and D. G. Cahill, "Heat transport in thin dielectric films," *J. Appl. Phys.* **81**(6), pp. 2590–2595, 1997.

[16] J. Y. Duquesne, D. Fournier and C. Frétigny, "Analytical solutions of the heat diffusion equation for 3ω method geometry," *J. Appl. Phys.* **108**(8), p. 086104, 2010.

[17] J. H. Lienhard IV and J. H. Lienhard V, *A Heat Transfer Textbook*, 4th ed. Cambridge, MA: Phlogiston Press, 2012.

[18] Z. Chen, W. Jang, W. Bao, C. N. Lau and C. Dames, "Thermal contact resistance between graphene and silicon dioxide," *Appl. Phys. Lett.* **95**(16), pp. 161910–161913, 2009.

[19] L. Shi, "Mesoscopic thermophysical measurements of microstructures and carbon nanotubes," PhD dissertation, University of California at Berkeley, 2001.

[20] P. Kim, L. Shi, A. Majumdar and P. McEuen, "Thermal transport measurements of individual multiwalled nanotubes," *Phys. Rev. Lett.* **87**(21), p. 215502, 2001.

[21] D. Li, "Thermal transport in individual nanowires and nanotubes," PhD dissertation, University of California at Berkeley, 2002.

[22] J. H. Seol, I. Jo, A. L. Moore, L. Lindsay, Z. H. Aitken, M. T. Pettes, X. Li, Z. Yao, R. Huang, D. Broido, N. Mingo, R. S. Ruoff and L. Shi, "Two-dimensional phonon transport in supported graphene," *Science* **328**(5975), pp. 213–216, 2010.

[23] L. Shi, D. Y. Li, C. H. Yu, W. Y. Jang, D. Kim, Z. Yao, P. Kim and A. Majumdar, "Measuring thermal and thermoelectric properties of one-dimensional nanostructures using a microfabricated device," *J. Heat Transfer Asme* **125**(5), pp. 881–888, 2003.

[24] J. Tang, H.-T. Wang, D. H. Lee, M. Fardy, Z. Huo, T. P. Russell and P. Yang, "Holey silicon as an efficient thermoelectric material," *Nano Lett.* **10**(10), pp. 4279–4283, 2010.

[25] M. T. Pettes and L. Shi, "Thermal and structural characterizations of individual single-, double-, and multi-walled carbon nanotubes," *Adv. Funct. Mater.* **19**(24), pp. 3918–3925, 2009.

[26] K. Hippalgaonkar, B. Huang, R. Chen, K. Sawyer, P. Ercius and A. Majumdar, "Fabrication of microdevices with integrated nanowires for investigating low-dimensional phonon transport," *Nano Lett.* **10**, pp. 4341–4348, 2010.

[27] A. Weathers and L. Shi, "Thermal transport measurement techniques for nanowires and nanotubes," *Annu. Rev. Heat Transf.* **16**(1), pp. 101–134, 2013.

[28] M. Fujii, X. Zhang, H. Xie, H. Ago, K. Takahashi, T. Ikuta, H. Abe and T. Shimizu, "Measuring the thermal conductivity of a single carbon nanotube," *Phys. Rev. Lett.* **95**(6), p. 065502, 2001.

[29] C. Dames, S. Chen, C. T. Harris, J. Y. Huang, Z. F. Ren, M. S. Dresselhaus and G. Chen, "A hot-wire probe for thermal measurements of nanowires and nanotubes inside a transmission electron microscope," *Rev. Sci. Instrum.* **78**(10), p. 104903, 2007.

[30] W. Jang, W. Bao, L. Jing, C. N. Lau and C. Dames, "Thermal conductivity of suspended few-layer graphene by a modified T-bridge method," *Appl. Phys. Lett.* **103**, p. 133102, 2013.

[31] J. Kim, D. Seo, H. Park, H. Kim, H. Choi and W. Kim, "Extension of the T-bridge method for measuring the thermal conductivity of two-dimensional materials," *Rev. Sci. Instrum.* **88**(5), p. 054902, 2017.

[32] W. Jang, Z. Chen, W. Bao, C. N. Lau and C. Dames, "Thickness-dependent thermal conductivity of encased graphene and ultrathin graphite," *Nano Lett.* **10**(10), pp. 3909–3913, 2010.

[33] L. Lu, W. Yi and D. L. Zhang, "3ω method for specific heat and thermal conductivity measurements," *Rev. Sci. Instrum.* **72**(7), pp. 2996–3003, 2001.

[34] B. K. Park, J. Park and D. Kim, "Note: Three-omega method to measure thermal properties of subnanoliter liquid samples," *Rev. Sci. Instrum.* **81**(6), p. 066104, 2010.

[35] F. Chen, J. Shulman, Y. Xue, C. W. Chu and G. S. Nolas, "Thermal conductivity measurement under hydrostatic pressure using the 3ω method," *Rev. Sci. Instrum.* **75**(11), pp. 4578–4584, 2004.

[36] I. K. Moon and Y. H. Jeong, "The 3ω technique for measuring dynamic specific heat and thermal conductivity of a liquid or solid," *Rev. Sci. Instrum.* **67**(1), pp. 29–35, 1996.

[37] D. Oh, A. Jain, J. K. Eaton, K. E. Goodson and J. S. Lee, "Thermal conductivity measurement and sedimentation detection of aluminum oxide nanofluids by using the 3ω method," *Int. J. Heat and Fluid Flow* **29**(5), pp. 1456–1461, 2008.

[38] S. D. Lubner, J. Choi, G. Wehmeyer, B. Waag, V. Mishra, H. Natesan, J. C. Bischof and C. Dames, "Reusable bi-directional 3ω sensor to measure thermal conductivity of 100-μm thick biological tissues," *Rev. Sci. Instrum.* **86**, p. 014905, 2015.

[39] B. K. Park, N. Yi, J. Park and D. Kim, "Thermal conductivity of single biological cells and relation with cell viability," *Appl. Phys. Lett.* **102**(20), p. 203702, 2013.

Chapter 3

How to Prepare for a Successful Experiment?

"The general who wins the battle makes many calculations before the battle; the general who loses makes but few calculations beforehand. Thus do many calculations lead to victory, and few calculations to defeat: how much more no calculation at all!"

— Sun Tzu, The Art of War

3.1. Introduction

In this chapter, we present various aspects of the experimental process which are usually omitted from a journal paper but which experience has shown are important for conducting a successful experiment. This includes thermal design to minimize parasitics, getting oriented through "pre-lab" estimates, and verification by control experiments and sanity checks. The issues are discussed in context of several of the techniques already introduced in earlier chapters.

3.2. Thermal design to make parasitics negligible

In an ideal experiment we would guide the heat to flow exclusively through the nanostructure of interest, but in reality there are various other series and parallel pathways. To be able to neglect those parasitics requires careful thermal design.

Using the heat spreader method of Secs. 1.8 and 2.3.2(F) as a concrete example (Fig. 3.1), we discuss the thermal design to ensure

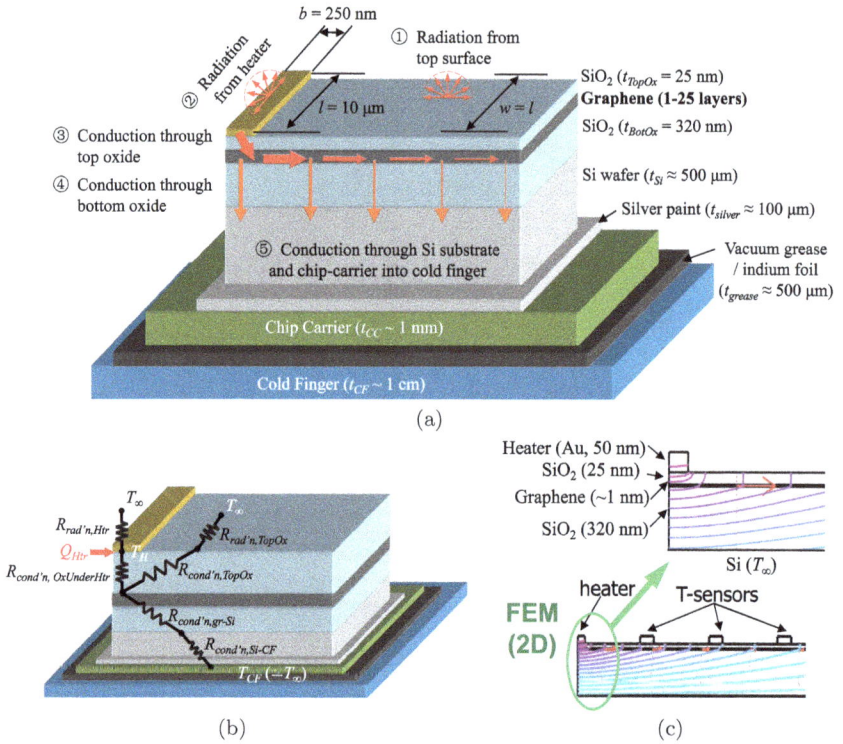

Fig. 3.1   Heat spreader method as an example to highlight some key considerations in a typical thermal design, which will be discussed in detail in Secs. 3.2.1 to 3.2.4. (a) Schematic of the sample, simplified from Fig. 1.8 using a symmetry plane. In addition, the chip carrier (CC) and the cold figure (CF) are also illustrated. Note that the μm-scale heater pattern is exaggerated to show the finest features. (b) An approximate lumped resistor-network overlaid on top of the cartoon in (a). Note that some of the thicknesses are distorted. (c) A more advanced 2D FEM analysis shows more details, e.g., the isothermal contours inside the top and bottom oxide layers. With its advantage of revealing physics in more intuitive algebraic forms, the analysis in this chapter is based on the approximate thermal circuit in (b).

negligible parasitics. We take advantage of the symmetry of the problem (Fig. 1.8) to analyze only half of the sample, as shown in Fig. 3.1(a). More details about how the sample is mounted to the cold finger of the cryostat through a chip carrier are also sketched in Fig. 3.1(a). Here to depict the finest features on the same cartoon we have greatly exaggerated the size of the μm-scale heater pattern and thin film thicknesses.

Recall the basics from Sec. 1.8. The metallic line heater generates Joule heat at a rate Q_{Htr}, which flows vertically through the top oxide and then spreads laterally through the high-k graphene layer, while simultaneously leaking vertically through the bottom oxide layer, and finally dissipates through the Si substrate and chip carrier into the cold finger which is set at a specific temperature ($T_{\text{CF}} \approx T_\infty$) for measurement.

A corresponding thermal circuit describing the physics above is shown in Fig. 3.1(b). The key resistor of interest is approximated as a fin,

$$R_{\text{cond'n,gr–Si}} = \frac{1}{\sqrt{h_{\text{eff}}P k_{\text{gr},\|} A_{\text{cr}}}}, \tag{3.1}$$

with a characteristic fin length

$$\beta^{-1} = \sqrt{k_{\text{gr},\|} A_{\text{cr}} / h_{\text{eff}} P}, \tag{3.2}$$

in which $k_{\text{gr},\|}$ is the thermal conductivity, and A_{cr} and P are the cross-sectional area and "wetted perimeter" of the graphene flake. For a flake of width w and thickness t_{gr}, we have $A_{\text{cr}} = w \cdot t_{\text{gr}}$ and $P = w$ because the flake only conducts heat out through its lower face. Recalling Eq. (2.34), we have the effective "convection" coefficient,

$$h_{\text{eff}} = (t_{\text{BotOx}}/k_{\text{BotOx}} + R''_{c,\text{gr–ox}} + R''_{c,\text{ox–Si}})^{-1}, \tag{3.3}$$

where k_{BotOx} and t_{BotOx} are the thermal conductivity and thickness of the lower oxide layer, and the two R''_c terms are the specific contact resistances (with SI units $\text{m}^2 \cdot \text{K/W}$) from graphene-to-oxide and from oxide-to-silicon, respectively. Evaluating Eq. (3.3) using the parameters from Table 3.1, we find $h_{\text{eff}} = 4.0 \times 10^6 \text{ W/m}^2 \cdot \text{K}$. Likewise, evaluating Eq. (3.2), we find $\beta^{-1} = 94$ nm and 1.3 μm for a bi-layer and a 21-layer graphene flake, respectively.

Ideally, we desire the thermal circuit to only contain $R_{\text{cond'n,gr–Si}}$. However, as indicated in Fig. 3.1(b) the practical thermal circuit also includes the following nonideal resistors: $R_{\text{rad'n,Htr}}$ and $R_{\text{rad'n,TopOx}}$

Table 3.1   Geometric parameters and thermal properties used in Sec. 3.2 for the example of the heat spreader method (Fig. 3.1) [1].

Parameters		Typical value	Units
Thickness of top and bottom	t_{TopOx}	0.025	
oxide layers	t_{BotOx}	0.32	μm
Length and half width of	l	10	
heater	b	0.25	
Width of graphene flake	w	10	
Thickness of graphene flake	t_{gr}	0.69 (bi-layer)	nm
		7.25 (21-layer)	
Thermal conductivities of top	k_{TopOx}	0.95	
and bottom oxide layers	k_{BotOx}	1.43	W/m$\cdot$K
Thermal conductivity of	$k_{\mathrm{gr},\parallel}$	51 (bi-layer)	
graphene flakes		970 (21-layer)	
TBR between gr. and ox,	$R''_{c,\mathrm{gr-ox}}$	9.5	10^{-9} m$^2 \cdot$
and between ox and Si	$R''_{c,\mathrm{ox-Si}}$	20	K/W
Characteristic fin length	β^{-1}	0.094 (bi-layer)	μm
calculated using Eq. (3.2)		1.3 (21-layer)	

correspond to radiation losses from the top surfaces of the line heater and sample surface, respectively; $R_{\mathrm{cond'n,OxUnderHtr}}$ is conduction through the portion of the top oxide which is underneath the line heater; similarly $R_{\mathrm{cond'n,TopOx}}$ represents the path through the top oxide above the graphene flake to the exposed free surface; and $R_{\mathrm{cond'n,Si-CF}}$ represents the thermal pathway from the Si substrate to the cold finger.

Although this lumped resistor network is a major simplification, it offers important physical insights to guide the first stage of thermal design. On the other hand, more advanced analysis offers more details and more accurate results, but at the loss of simplicity in expressing the physics. For example, a 2D FEM analysis assuming a perfect Si heat sink is shown in Fig. 3.1(c), which shows clearly the temperature distribution inside the layers. However, this advanced FEM analysis is based on a numerical scheme with 1000s of nodes, and cannot show the physics in more intuitive algebraic forms. Thus, our analysis in the following will be based on the lumped thermal circuit in Fig. 3.1(b).

To ensure that the Joule heat Q_{Htr} is most sensitive to $R_{\mathrm{cond'n,gr-Si}}$ which includes the thermal conductivity of the graphene flake (see

Eq. (3.1)), in the following we will test each of the nonideal thermal resistors against $R_{\text{cond'n,gr-Si}}$ one by one: the radiation losses in Sec. 3.2.1, the effects of the thermal resistance of the top oxide in Sec. 3.2.2, and the design of the metal heater pattern and the thermal contact to the cold finger in Sec. 3.2.3. Each such test focuses on a single nonideality and ignores the others. This approach is reasonable and self-consistent because only once all nonidealities are shown negligible can the entire design be considered a success. We also discuss the cross-cutting topic of designing experiments to be sensitive to the expected range of the measurand in Sec. 3.2.4.

3.2.1. *Radiation losses*

Here we assume the measurement is conducted in a standard cryostat at high vacuum ($<10^{-5}$ Torr), thus making negligible the heat losses through convection and air conduction. We will revisit the question of convection losses and vacuum level in Appendices B and E.

As indicated by labels 1 and 2 in Fig. 3.1, we consider radiation losses from the top surface of the sample and from the micro-fabricated heater. The goal is to ensure that nearly all of the Joule heat dissipated in the heater line is conducted downward into the SiO_2/graphene/SiO_2/Si sample stack, with negligible losses by radiation.

We first check the radiation loss by the top surface of the sample (label 1 in Fig. 3.1). To ensure it is small compared to the conduction through the sample stack, we require

$$R_{\text{rad'n,TopOx}} \gg R_{\text{cond'n,gr-Si}}, \tag{3.4}$$

in which

$$R_{\text{rad'n,TopOx}} = \frac{1}{h_{\text{rad'n}}A_{\text{surf}}}, \tag{3.5}$$

where A_{surf} is the effective top surface area of the sample, and $h_{\text{rad'n}}$ is a linearized radiative heat transfer coefficient [2, p. 254]

$$h_{\text{rad'n}} = 4\varepsilon\sigma T_{\text{avg}}^3, \tag{3.6}$$

in which ε is the emissivity of the sample, $\sigma = 5.67 \times 10^{-8}$ Wm^{-2}K^{-4} is the Stefan–Boltzmann constant, and T_{avg} the average of the temperatures of the sample and the ambient. For a black surface at room temperature, $h_{\text{rad'n}} \approx 6.1$ W/m$^2 \cdot$ K.

We evaluate the effective top surface area of the sample as

$$A_{\text{surf}} = w \cdot \beta^{-1}, \tag{3.7}$$

which is well justified since β^{-1} from Eq. (3.2) can be loosely understood as the characteristic in-plane distance over which the graphene heat spreader "feels" the effect of the heater.

Substituting Eqs. (3.1) and (3.5) into (3.4), the criteria to neglect the radiative loss from the top surface of the sample becomes

$$\frac{h_{\text{eff}}}{h_{\text{rad'n}}} \gg 1, \tag{3.8}$$

which is extremely well satisfied based on the estimates above, with $h_{\text{rad'n}} \approx 6.1$ W/m$^2 \cdot$ K and $h_{\text{eff}} \approx 3.9 \times 10^6$ W/m$^2 \cdot$ K.

The analysis above assumes

$$R_{\text{rad'n,TopOx}} \gg R_{\text{cond'n,TopOx}}, \tag{3.9}$$

and

$$R_{\text{con'n,gr–Si}} \gg R_{\text{cond'n,Si–CF}}. \tag{3.10}$$

We will confirm Eq. (3.10) in Sec. 3.2.4, and Eq. (3.9) now. We have

$$\frac{R_{\text{rad'n,TopOx}}}{R_{\text{cond'n,TopOx}}} = \frac{1/h_{\text{rad'n}}}{t_{\text{TopOx}}/k_{\text{TopOx}}}, \tag{3.11}$$

where both resistors involve the same A_{surf} which has therefore canceled out. Using the parameters from Table 3.1, we find this ratio to be $\sim 7.7 \times 10^6$, thereby confirming Eq. (3.9).

We next check the radiation loss by the microfabricated heater (label 2 in Fig. 3.1(a)). As shown in Fig. 3.1(b), to ensure it is small compared to the conduction through the sample stack, we require

$$R_{\text{rad'n,Htr}} \gg R_{\text{cond'n,gr-Si}}. \tag{3.12}$$

Using the linearized radiative heat transfer coefficient (Eq. (3.6)), we have

$$R_{\text{rad'n,Htr}} = \frac{1}{h_{\text{rad'n}}(l \cdot b)}. \tag{3.13}$$

Combining Eqs. (3.1) and (3.13), we have

$$\frac{R_{\text{rad'n,Htr}}}{R_{\text{cond'n,gr-Si}}} = \frac{w\sqrt{h_{\text{eff}}k_{\text{gr},\|}t_{\text{gr}}}}{h_{\text{rad'n}}(l \cdot b)}. \tag{3.14}$$

Using the bi-layer graphene sample as a conservative estimate, and the parameters from Table 3.1, we arrive at $R_{\text{rad'n,Htr}}/R_{\text{cond'n,gr-Si}} \simeq 2.4 \times 10^5$ even for a black heater surface ($\varepsilon_{\text{Htr}} = 1$), which satisfies Eq. (3.12) extremely well. This analysis also assumes $R_{\text{cond'n,OxUnderHtr}} \ll R_{\text{cond'n,gr-Si}}$, which we will confirm in Sec. 3.2.2.

The analysis in Sec. 3.2.1 points towards a more general conclusion, that at room temperature and below, radiation losses in most cases are negligible for micro- and nanoscale thermal measurements. This can be seen by comparing the room temperature $h_{\text{rad'n}} \approx 6.1$ W/m$^2 \cdot$ K to a conductive heater transfer coefficient

$$h_{\text{cond'n}} = \frac{k}{L_{\text{char}}}, \tag{3.15}$$

where L_{char} is a characteristic length of the sample. Even for a thermal insulator such as fused SiO$_2$ ($k \sim 1$ W/m $\cdot$ K), and a very large $L_{\text{char}} \approx 1$ mm, this $h_{\text{cond'n}}$ is still on the order of 1000 W/m$^2 \cdot$ K, which is more than two orders of magnitude higher than $h_{\text{rad'n}}$.

For high temperature (e.g., up to 1000 K) experiments however, the radiation loss may not be neglected since $h_{\text{rad'n}} \propto T_{\text{avg}}^3$ (Eq. (3.6)). More discussion can be found in Appendix F.

3.2.2. *Dielectric isolation and its thermal resistance*

In many of the electrothermal techniques, we microfabricate metal electrodes on top of the sample, flow current through them, and measure the corresponding voltages across the electrodes. In these scenarios, any leakage of the electrical current from the metal electrodes to the sample may ruin the measurement. Thus, it is crucial to ensure electrical insulation between the electrodes and the sample.

One good example is the heat spreader method. As indicated by label 3 in Fig. 3.1(a), because graphene is electrically conducting, we cannot directly fabricate the metal electrodes on top of it, but instead require an insulating layer between the electrodes and graphene. Unfortunately, as shown in Fig. 3.1(b), this electrically insulating layer also acts as an additional thermal resistor ($R_{\text{cond'n,OxUnderHtr}}$) in series between the metal heater line and the graphene heat spreader. If knowing the temperature of the heater line is important for the experimental determination of graphene's k, we require this dielectric series resistance to be negligible as compared to the graphene heat spreading resistor:

$$R_{\text{cond'n,OxUnderHtr}} \ll R_{\text{cond'n,gr-Si}}. \tag{3.16}$$

Assuming 1D heat transfer through the portion of the top oxide underneath the line heater, a treatment supported by the approximately parallel isotherms in that location in the FEM simulations of Fig. 3.1(c), we have

$$R_{\text{cond'n,OxUnderHtr}} = \frac{t_{\text{TopOx}}}{k_{\text{TopOx}}(b \cdot l)}. \tag{3.17}$$

Combining Eq. (3.1) and (3.17), we have

$$\frac{R_{\text{cond'n,OxUnderHtr}}}{R_{\text{cond'n,gr-Si}}} = \frac{wt_{\text{TopOx}}}{l \cdot b} \sqrt{\frac{k_{\text{gr},\parallel}}{k_{\text{BotOx}}} \frac{t_{\text{gr}}}{t_{\text{BotOx}}}}. \tag{3.18}$$

Using the parameters from Table 3.1, we find $R_{\text{rad'n,OxUnderHtr}}/R_{\text{cond'n,gr-Si}} \simeq 0.03$ for a bi-layer graphene sample, which satisfies Eq. (3.16). On the other hand, for the 21-layer graphene flake we estimate $R_{\text{cond'n,OxUnderHtr}}/R_{\text{cond'n,gr-Si}} \simeq 0.5$, showing that $R_{\text{cond'n,OxUnderHtr}}$ cannot always be neglected. As a result, the actual model used in [1] excluded the temperature of the heater from the fitting.

3.2.3. *Design of thermal contacts and heater pattern: Minimizing artifacts of unintended background heating*

From the point of view of electrical measurements, four-probe measurements have the great advantage of being immune to the effects of lead resistance (see Appendix C). From the point of view of thermal measurements, however, the current-carrying leads (conventionally designated I^+ and I^-) still cause Joule heating, so some additional care is required to design the pattern of the heater and the thermal contacts properly when applying four-probe resistance thermometry for a thermal measurement.

A representative example based on our experience with a heat spreader method [1] is shown in Fig. 3.2. Our initial, and in hindsight naïve, design is shown in Figs. 3.2(a) and 3.2(c). Note in Fig. 3.2(a) that the I^+ and I^- leads are thin and very long: a few mm. These current leads result in substantial extra resistances in series with the central portion of the heater between the V^+ and V^- taps (each $\sim 0.5\ \mu$m wide, on either side of the sample of width $l \approx 10\ \mu$m as shown in Fig. 3.1(a)). These extra resistances of the long current leads cause a background of undesired Joule heating, the magnitude of which greatly exceeds the desired Joule heating in the $10\ \mu$m test section between the V^+ and V^- taps. This background heating must dissipate out to the heat sink at T_∞, and in doing so may cause substantial DC heating of the test section for reasons that have nothing to do with the thermal properties of the graphene itself.

As indicated by label 5 in Fig. 3.1(a), this problem is amplified if the thermal contacts of the sample mounting inside the cryostat are not well designed. Figure 3.2(c) shows a typical configuration of a thermal measurement conducted inside a cryostat, and Fig. 3.2(e) shows its corresponding thermal circuit. The DC heating must flow out to the bottom of the substrate (a silicon wafer), through silver paint into a ceramic chip carrier, and finally through a thin layer of high vacuum grease (Dow Corning) into a temperature controlled copper cold finger. Estimates for each of these thermal resistances are given in Table 3.2, which shows that the dominant resistance

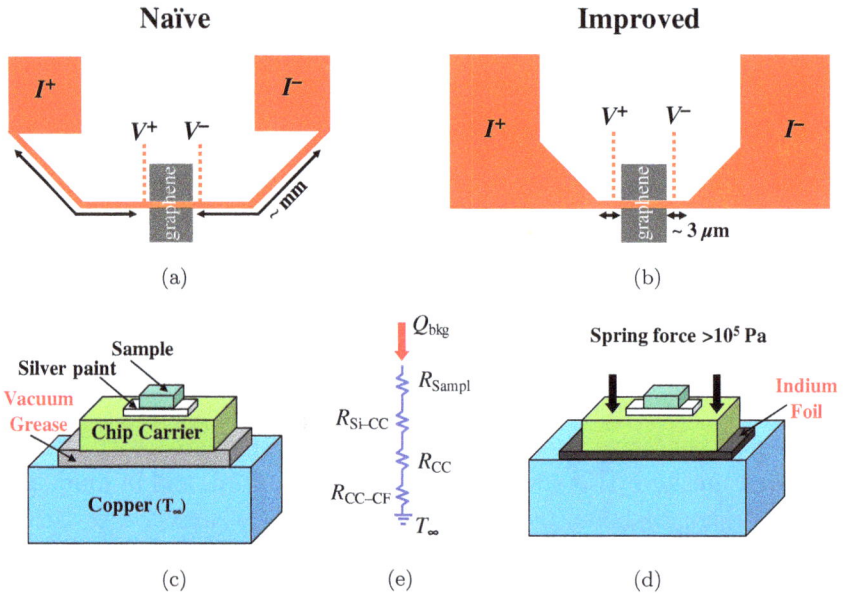

Fig. 3.2 Naïve (left column) vs. improved (right column) designs for line-heater and thermal contact in the heat spreader method of Fig. 3.1. Top row: top views of the heater metallization, with typical line widths 0.5 μm. Bottom row: schematic 3D view at low magnification, showing the thermal contact among sample (graphene on Si chip), ceramic chip carrier (CC), and copper cold finger (CF) at T_∞. (a) Long and narrow current leads cause undesired Joule heating far from the graphene specimen, and thus a large background temperature rise which confuses interpretation of the local temperature measurements at the graphene. (b) Improved design uses short and wide current leads to minimize this background heating. (c) Poor thermal contact relying on vacuum grease between chip carrier and copper cold finger of a cryostat dominates the thermal circuit (see Table 3.2), and amplifies the temperature rise caused by the excessive background heating from (a). (d) Improved design replaces the vacuum grease with thin indium foil and applies clamping pressure, both of which reduce thermal contact resistance between chip carrier and copper cold finger. Combining the improvements from (b) and (d), the temperature rise artifact at the graphene due to background heating is reduced by at least two orders of magnitude. (e) Lumped thermal circuit to show the heat transfer path from the sample to the copper cold finger. The difference between (c) and (d) is represented in the thermal contact resistance between the chip carrier and the copper cold finger, R_{CC-CF}.

Table 3.2   Estimating the thermal resistances between sample and copper cold finger for a typical experiment inside a benchtop cryostat (Figs. 3.2(c)–3.2(d)). Note that here we crudely assume locally 1D heat transfer through the stack, and use an effective area ratio $A_{sampl}/A_{Carrier} \approx 0.5$ to roughly take into account the spreading effect when heat flows from a small size sample to a large size chip carrier and cold finger.

Labels in Fig. 3.2	Note	Thermal k (W/m·K)	Area (cm²)	Thickness (μm)	Thermal resistance (K/W)
R_{Si-CC}	Silver Paint	8	1	100	0.10
R_{CC}	Chip Carrier (ceramic)	17	1	3000	1.80
$R_{CC-CF,naive}$	Vacuum Grease	0.2	2	500–1000	**13–25**
$R_{CC-CF,improved}$	Indium Foil	80	2	500	**0.06**

is the vacuum grease layer between the chip carrier and copper sample stage. Coupled with the unnecessarily large background heating caused by the long thin current leads, this results in a "background" temperature rise which is spatially uniform on the scale of the graphene test sample. As we shall see next, this led to large measurement errors when comparing the experimental results with a thermal model which considered only the local heating by the central 10 μm portion of the heater between the V^+ and V^- taps.

Figure 3.3 shows more detail about these two effects. In order to verify the agreement between the measurement and the 3D FEM model in the heat spreader method, we performed a control experiment to measure the temperature response on the top surface of the top oxide for 9 distances ranging from 1 to 1000 μm away from the heater line. Note that this control experiment has no graphene layer, and thus the FEM has no fitting parameters. As shown schematically in Fig. 3.2(a), our initial design of the heater line had very long I^+ and I^- leads, which dissipated over 20 times more heating power than the central portion of the heater! Comparing curves (1) and (2) in Fig. 3.3, including this far-field Joule heating in the 3D FEM model helps bring the model closer to the experimental

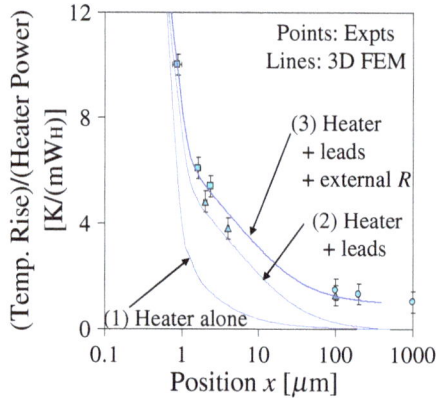

Fig. 3.3   An example of a large effect of DC background heating in a sample with long I^+/I^- leads, expressed as temperature rise per mW of power at the central heater line. This control experiment used a sample similar to Fig. 3.1 but excluding the graphene layer, and measured temperature at points spanning a larger range of distances from the heater line. For this naïve design as shown in Figs. 3.2(a) and 3.2(c), in order to reconcile the measurements with the 3D FEM simulations, the FEM model must include the extra Joule heating from the long heater leads as well as the external thermal resistances dominated by the grease layer (see the progression of model curves $1 \rightarrow 2 \rightarrow 3$). These results suggest two improvements as indicated in Figs. 3.2(b) and 3.2(d).

points. However, this requires a much larger simulation domain which increases the computational time, and there are still major discrepancies in the temperature field especially for larger x.

In addition, as shown in Fig. 3.2(c), our original design relied on a layer of vacuum grease ($k \sim 0.2$ W/m·K) to make thermal contact between the chip carrier and copper sample stage. We estimate that the thermal resistance of this layer is 13–25 K/W. Thus, for every 1 mW dissipated in the central heater line, another 20 mW is dissipated in the I^+/I^- leads, corresponding to 0.3–0.5 K of temperature drop through the grease, which accordingly increases the temperature of the sample measurement region by the same amount. Thus, we estimate that the combined effects of Joule heating in the current leads and poor heat sinking between sample and copper stage are enough to cause a significant uniform "background" temperature rise of ~ 0.5 K/m W_{Htr}, which affects all T sensors uniformly (see curve 3 of Fig. 3.3).

Based on the lessons learned in Fig. 3.3, we improved the experiment:

- The heater current leads were made much shorter and wider (see the schematic in Fig. 3.2(b) and the SEM images in Fig. 3.4(b)), reducing the background heating by a factor of at least 10.
- We replaced the grease with indium foil ($k \sim 80$ W/m·K) and built a simple spring fixture to apply several atmospheres of clamping pressure (see the schematic in Fig. 3.2(d)). Referring to the estimates of Table 3.1, this change should reduce the total thermal resistance for heat sinking, and thus background temperature rise, by another factor of at least 10.
- Recognizing that the three-dimensional aspect of the heat spreading was important for the overall thermal response, in the improved design we trimmed the graphene flake from an irregular (Fig. 3.4(a)) to a rectangular shape (Fig. 3.4(b)).

3.2.4. *Identifying a technique's best sensitivity range*

Every experimental technique is best suited for measuring only a limited range of property values, depending on sample's dimensionality/

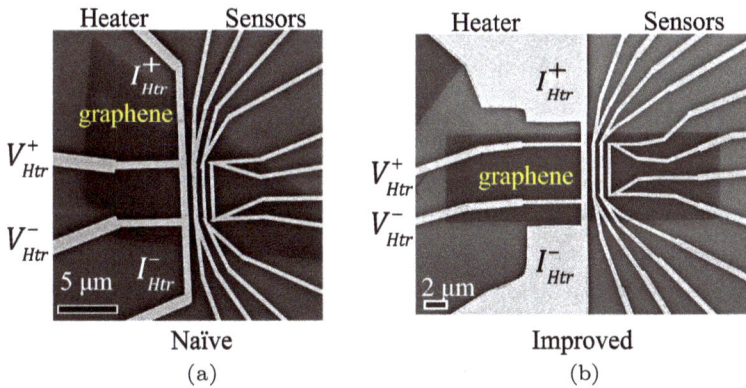

Fig. 3.4 Top view SEM images for (a) original design with long and narrow lines for the heater's I^+ and I^- leads, and (b) improved design with short and fat leads. These images correspond to Figs. 3.2(a) and 3.2(b). These SEM images also show that the graphene flake has been trimmed to a rectangular shape in the improved design, a more regular geometry which is also helpful for the analysis.

geometry and thermal properties. We have discussed the former in detail in Chapter 2. We now discuss the latter, continuing the case study of the heat spreader method.

Although this heat spreader method was developed for measuring multilayer graphene, it turns out that it was not sensitive to the thinnest of films, e.g., single-layer graphene with relatively low thermal conductivity. This limitation arises from the dual functions of the bottom oxide layer, as shown in Fig. 3.1(a). Besides its role in the thermal model, this oxide is critical for determining the number of the graphene layers by an optical interference method, which requires its thickness to be no less than $\sim$300 nm [3]. We shall now see how such relatively thick oxide limits the sensitivity of the heat spreader method for ultra-thin films with low thermal conductivity.

This limitation can be qualitatively understood from the requirement of the fin model (see Sec. 2.3.2(F) for details), which requires

$$\beta^{-1} \geq t_{\text{BotOx}}, \tag{3.19}$$

where β^{-1} is the fin length. Recalling Eq. (3.2)

$$\beta^{-1} = \sqrt{k_{\text{gr},\|} A_{\text{cr}} / h_{\text{eff}} P},$$

and approximating h_{eff} from Eq. (3.3) to be $k_{\text{BotOx}}/t_{\text{BotOx}}$, since the two contact resistance are orders of magnitude smaller than $t_{\text{BotOx}}/k_{\text{BotOx}}$, we simplify Eq. (3.19) as

$$(k \cdot t)_{\text{gr}} \geq (k \cdot t)_{\text{BotOx}}, \tag{3.20}$$

which suggests that the heat spreader method, if fitting to a 1D fin model, will lose its sensitivity to graphene samples with $(k \cdot t)_{\text{gr}}$ much smaller than $(k \cdot t)_{\text{BotOx}}$. Plugging in numbers, the bottom oxide layer has $(k \cdot t)_{\text{BotOx}} \approx 450$ [(W/m·K) × (nm)] while the bi-layer sample has $(k \cdot t)_{\text{BLG}} \approx 30$ [(W/m·K) × (nm)]. Thus, Eq. (3.20) shows that the 1D fin treatment is not appropriate for the thinnest samples.

Generalizing the data processing from a 1D fin approximation to a full 3D FEM treatment pushed the sensitivity limits to substantially smaller $(k \cdot t)_{\text{gr}}$ of $\sim$50 [(W/m·K) × (nm)]. Indeed, for the single-layer graphene (SLG) sample there was *no* measurable temperature rise at the three T sensors above the noise floor, which means that this sensitivity estimate establishes the upper bound on the SLG sample's thermal conductivity [1].

3.2.5. *Overall uncertainty of the heat spreader method*

Accounting for all of the various phenomena described above, for our specific geometries and properties of the stack, we evaluated the total uncertainty of this technique using a Monte Carlo method (Sec. 4.4).

Figure 3.5 shows the uncertainty as a function of the product of the thermal conductivity and the thickness $(k_{gr,\parallel} \times t_{gr})$ of graphene and ultra-thin graphite thin films [1]. Note that the uncertainty is defined as the ratio of the half of the calculated 95% confidence interval $(k_{LB} < k < k_{UB})$ to the experimental

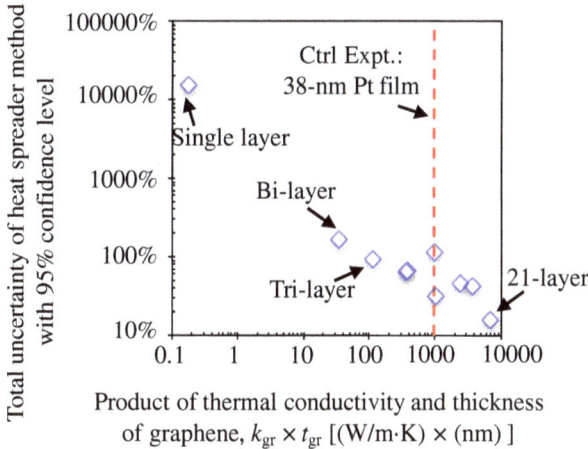

Fig. 3.5 Total uncertainty of the heat spreader method as a function of the product of the thermal conductivity and the thickness, $(k_{gr,\parallel} \times t_{gr})$, in a log–log scale. This uncertainty is defined as the ratio of half of the calculated 95% confidence interval to the experimental result, $\frac{1}{2}(k_{gr,\parallel,UB} - k_{gr,\parallel,LB})/k_{gr,\parallel,expt}$. The symbols represent measurements on graphene samples with different thicknesses, while the dashed vertical line indicates the $[k \times t]$ value of the control experiment on a Pt thin film with a thickness of 38 nm. The dominant feature is that the uncertainty decreases with increasing $(k_{gr,\parallel} \times t_{gr})$. The measurements of the thinnest samples, especially the SLG, were only meaningful for establishing an upper bound on k_{Gr}, so that the corresponding calculated uncertainty appears absurd.

result, $\frac{1}{2}(k_{\mathrm{gr},\|,\mathrm{UB}} - k_{\mathrm{gr},\|,\mathrm{LB}})/k_{\mathrm{gr},\|,\mathrm{expt}}$. The blue symbols represent our measurements on different samples, ranging from a single-layer sample to a 21-layer sample. The red vertical dashed line indicates a control experiment to validate the spreader method, in which we extracted the thermal conductivity of a 38-nm-thick Pt thin film using this technique and compared to a separate experiment combining resistance thermometry and the Wiedemann–Franz law (see more details in Sec. 3.4 and [1]), which demonstrated an agreement of better than 5%.

The key feature of Fig. 3.5 is that the heat spreader method loses its accuracy as $(k_{\mathrm{gr},\|} \times t_{\mathrm{gr}})$ decreases. For example, the uncertainty for the 21-layer thick sample at 95% confidence interval is ±16% (i.e., $\frac{1}{2}(k_{\mathrm{gr},\|,\mathrm{UB}} - k_{\mathrm{gr},\|,\mathrm{LB}})/k_{\mathrm{gr},\|,\mathrm{expt}} = 0.16$). This uncertainty increases to >10000% for the single-layer sample, an absurd value indicating that the measurement in fact is no longer able to detect the actual $k_{\mathrm{gr},\|}$. This does not mean the measurement is empty of meaning, however; rather, as noted in the previous subsection, this measurement still had value in establishing an (unexpectedly low) upper bound on the SLG thermal conductivity [1].

More details of this uncertainty analysis can be found in Table 4.1 and Sec. 4.4.

3.3. "Pre-lab exercises": Estimating key electrical parameters prior to the experiment

Similar to the various thermal estimates presented above, it is also recommended to estimate some of the key electrical parameters expected in an experiment prior to taking any data. Among other things, this helps ensure that the appropriate equipment is selected. Further, gross errors are more quickly noticed if the experimentalist already has an idea of the ballpark values of expected signals. In the following two examples, we outline some of these key estimates in the context of the classic 3ω method.

Example 3.1. Consider a 3ω method to measure the thermal conductivity of an undoped Si wafer at room temperature, as presented in Sec. 2.2.1. Making reasonable assumptions as needed, give numerical estimates for:

(a) How much Joule heating is required?
(b) What electric current should we apply?
(c) What 3ω voltage is anticipated?

Solution: This problem statement is underspecified, and a number of additional assumptions are needed to complete the estimates. As shown in the flowchart of Fig. 3.6, before we can even estimate how much Joule heating is needed, we will need to specify a target $\Delta T_{\mathrm{sub},x}$, the amplitude of the oscillating part of the temperature response. For a measurement in a temperature environment at $T_\infty \approx T_{\mathrm{thermal\text{-}stage}}$, where $T_{\mathrm{thermal\text{-}stage}}$ is the set point of a temperature controller in K, usually we want $\Delta T < 0.01 T_{\mathrm{thermal\text{-}stage}}$, so that there is minimal ambiguity in reporting the measured k as a property measured simply at $T_{\mathrm{thermal\text{-}stage}}$. On the other hand, small $\Delta T_{\mathrm{sub},x}$ also makes all the signals smaller, which is more challenging from a signal-to-noise perspective. Here as a compromise let us choose $\Delta T_{\mathrm{sub},x} = 0.5$ K.

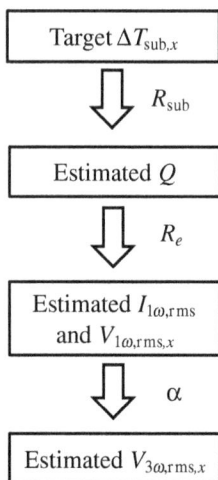

Fig. 3.6 Flowchart showing the procedure to estimate the magnitude of Joule heating, 1ω current, 1ω voltage, and 3ω voltage.

The next step in relating Q and $\Delta T_{\text{sub},x}$ is to estimate the real part of the thermal transfer function, R_{sub}. From Eq. (2.13), we have

$$R_{\text{sub}} = \frac{1}{\pi l k_{\text{sub}}} \left[\ln \left(\frac{L_p}{b} \right) + \eta \right], \tag{3.21}$$

where l and b are the length and half-width of the heater line, $\eta \approx 0.923$, L_p is the penetration depth, and k_{sub} is the thermal conductivity of the sample.

Note here R_{sub} is a function of the geometry of the heater, the thermal properties of the sample, and the heating frequency. We do not know any of those parameters, so let us consider them one by one to come up with reasonable estimates. Although the heat capacity of silicon is easily looked up, for rough estimates even this is not necessary, because as discussed in Appendix G.1 a reasonable estimate is $C \sim 3 \times 10^6$ J/m$^3 \cdot$ K for almost any material at room temperature and above. We should also have an order-of-magnitude estimate for k_{sub} based on our tentative understanding of the sample. In this example we are not given any specifications about doping or microstructure of this Si wafer. Based on the literature this suggests k_{sub} might range anywhere from $\sim$10 W/m$\cdot$K to 150 W/m$\cdot$K. Because for a given $\Delta T_{\text{sub},x}$, Q is directly proportional to k_{sub}, to ensure we reach $\Delta T_{\text{sub},x} = 0.5$ K for this entire range of k_{sub} the conservative choice is to complete the subsequent calculations using $k_{\text{sub}} = 150$ W/m$\cdot$K. (Although we do not do so here, it would also be prudent to repeat the calculations for the other extreme, $k_{\text{sub}} = 10$ W/m$\cdot$K, to establish both upper and lower bounds on the electrical signals.)

Here based on experience we use a typical value for the frequency of the driving current, 1000 Hz, leaving the detailed estimate of a proper frequency range for the next example problem. Similarly, based on experience, for the dimensions of the microfabricated heater line we choose typical values 1 mm in length and 10 μm in width. Plugging everything into Eq. (3.21), we estimate $R_{\text{sub}} = 8$ K/W. Thus, to reach $\Delta T_{\text{sub},x} = 0.5$ K we will require a Joule heating of around 60 mW.

We can now estimate the electrical resistance, R_e, of the metal heater line. Let us use Au as an example, and assume it is 300 nm thick. The handbook resistivity for pure bulk gold is $2.44 \times 10^{-8}\,\Omega\cdot m$ (however, this will be significantly increased in a real microfabricated film; see Appendix G.4). This gives an estimated $R_e \approx 8\,\Omega$.

With the Joule heating of 60 mW and the electrical resistance of $8\,\Omega$, we estimate the driving electric current to be $I_{1\omega,\mathrm{rms}} = \sqrt{Q/R_e} \approx 90$ mA, and from Ohm's law the 1ω voltage to be 720 mV.

Finally, using Eq. (2.12) with a temperature coefficient of resistance of a typical microfabricated Au line heater as $\alpha \approx 3.4 \times 10^{-3}\mathrm{K}^{-1}$ (handbook value; see also Appendix G.4), we estimate the 3ω voltage to be around 600 μV. Both the 1ω and the 3ω voltages are well within the range of a routine measurement, suggesting this should be a straightforward experiment once the bugs are worked out.

Example 3.2. Estimate appropriate frequency ranges of the driving current for a 3ω measurement of a 500-micron-thick undoped silicon wafer. Repeat for a 1-cm-thick fused silica sheet.

Solution: Here our approach is to determine the upper and lower bounds of the frequency range by comparing the penetration depth with the geometry of the heater and sample, in order to satisfy the basic assumptions of the heat transfer model.

Recall the physical picture (Fig. 2.1) and the expression of penetration depth (Eq. (2.1)), and rewrite it as

$$L_p = \sqrt{\frac{D_{\mathrm{sub}}}{4\pi f}}, \qquad (3.22)$$

where the Joule heating frequency, ω_H, is double that of the electrical current,

$$\omega_H = 2\omega = 4\pi f.$$

We also recall from Sec. 2.2 the two assumptions of the classic 3ω method. Firstly, an infinitely narrow heater:

$$L_p \gg b; \qquad (3.23)$$

and secondly, a semi-infinite sample:

$$L_p \ll t, \tag{3.24}$$

where b is the half width of the line heater and t is the thickness of the sample.

Combining Eqs. (3.22)–(3.24), we come up with a generic expression

$$f_m = \frac{D_{\text{sub}}}{4\pi(cL_G)^2}, \tag{3.25}$$

to link the extreme (subscript m, meaning either minimum or maximum) frequency of electrical current, f_m, and the corresponding geometry, L_G. Here $L_G = b$ in Eq. (3.23) and $L_G = d$ in Eq. (3.24). Note c is a factor of safety, a pure number corresponding to the "much larger" in Eq. (3.23), or the "much smaller" in Eq. (3.24). Quantitative results for the errors introduced by selected values of c have been collected in [5, Table 3].

Substituting a typical $b = 5\,\mu\text{m}$ for a heater line prepared by standard photolithography, $D_{\text{sub}} = 7.5 \times 10^{-5}\ \text{m}^2/\text{s}$ for undoped silicon at room temperature, and taking $c = 5$, we obtain the upper bound $f_{\text{max}} = 9600\,\text{Hz}$.

Likewise, substituting $d = 500\,\mu\text{m}$ for the thickness of a Si wafer and a safety factor $c = 1/5$, we obtain the lower bound $f_{\text{min}} = 600\,\text{Hz}$.

Following a similar procedure using the thermal diffusivity of fused silica at room temperature, $D_{\text{sub}} = 7.5 \times 10^{-7}\ \text{m}^2/\text{s}$, we estimate the proper frequency range for a 1-cm-thick silica sheet to be 0.015–95 Hz. For ultralow frequency measurements ($f < 0.1$ Hz), the long time constant required of the Lock-In Amplifier is also a practical concern (see Appendix A.3).

3.4.　Control experiments: Validation using samples of known k

After carefully designing their new experiment to be feasible based on extensive estimates such as presented above, the experimentalist is no doubt eager to finally get to work collecting data on real samples. But before measuring novel materials, it is wise to first validate a new experimental apparatus using control experiments on standard materials with well-known thermal properties. It is best practice to validate the rig over a wide range of property values, e.g., using two or more standard samples with a range of k bracketing the expected k of the unknown samples.

A good candidate material for low-k validation is amorphous SiO_2 (a-SiO_2; also known as fused silica). One key merit of bulk a-SiO_2 is that its thermal conductivity is independent of the characteristic lengths of the experiment, such as heater width and film thickness, down to 10s of nanometers [6, 7]. However, the thermal conductivity of thin films may depend on the growth type. For example, different a-SiO_2 thin films grown thermally [8], sputtered [8], evaporated [8], or PECVD [9], with thickness ranging from 10s of nanometers to several microns, have been measured and found that the thermal conductivity can be reduced to around half of the bulk value.

A good candidate for high-k validation at the 100 μm scale and larger is intrinsic silicon, which is also well studied [6, 10]. Doped samples are not as helpful, because silicon's k strongly depends on both doping level and species [6] and such samples are less well characterized in the literature. Also, the phonon mean free paths in silicon are quite long, so that the effective thermal conductivity may be significantly reduced in samples with characteristic lengths below $\sim$10 μm [6, 11, 12]. Thus, a convincing control experiment should be conducted on an intrinsic bulk silicon wafer, rather than any microfabricated thin-or thick-film (unless of course k of the film was also measured independently).

For thin films, high-k validation samples are much harder to come by. Fundamentally this is because high-k materials are generally crystalline and have relatively long mean free paths for both phonons and electrons (10–100 nm for electrons in metals [13], and can easily reach beyond $\sim$10 μm for phonons [14, 15]). This makes these materials sensitive to extrinsic scattering effects by defects, external surfaces, and internal grain boundaries, all of which are expected in a microfabricated sample, reducing k well below handbook values (Appendix G.4) but by an extent which is hard to predict accurately. In such cases, it is recommended to cross-check k of a validation sample using more than one apparatus or technique. For example, some labs have access to a laser-based metrology method such as time- or frequency-domain thermoreflectance (TDTR or FDTR) which are well-suited for measuring k of thin films. Also research groups may be open to sharing samples which they have previously measured and reported in the literature.

If metal films are an option for a validation sample, another approach takes advantage of the Wiedemann–Franz law, a remarkable linear relationship between the thermal conductivity (k_e) and electrical conductivity (σ) of metals:

$$k_e = L_0 \sigma T, \qquad (3.26)$$

where T is the absolute temperature in Kelvin and $L_0 = 2.44 \times 10^{-8} \mathrm{W} \cdot \Omega \cdot \mathrm{K}^{-2}$ is the Lorenz number. Crucially, for noncryogenic temperatures the Wiedemann–Franz law is largely unaffected by extrinsic scattering mechanisms such as defects, surfaces, and grain boundaries, so that the bulk handbook value of L_0 can be expected to still hold to within $\pm \sim 10\%$ in a microfabricated metal film [16–18], even though σ_{Film} may be smaller than $\sigma_{\mathrm{Handbook}}$ by a factor of 2 or more (Appendix G.4). We used this approach to help validate the heat spreader method for measuring the in-plane k of encased thin films (see Sec. 2.3.2(F) and Fig. 3.5) [1]. In a single evaporator run, we prepared two nominally-identical 38-nm-thick Pt thin film samples. We measured σ from one sample using a four-probe resistance measurement, which using Eq. (3.26) gives an estimate of k. The other sample was used to measure k directly using the heat spreader method. These two values of k agreed to within 5%, which helped increase our confidence to move ahead and use this new method to study the k of graphene samples.

3.5.   Electrical sanity checks: Repeatability, reversibility, and scaling

When commissioning a new electrothermal measurement setup, and for general debugging, we have found it helpful to exercise the electrothermal response of the experiment through a broader-than-normal parameter space to confirm certain symmetries and scaling trends. Such checks increase confidence that the instruments are measuring what we think they are. On the other hand, this requires taking extra data beyond what is actually needed to extract a thermal property, and thus are not normally part of day-to-day operations.

Extra care should be taken for DC measurements, since they are easily affected by nonidealities such as Seebeck voltages, low-frequency drifts, and power-line noise, as discussed in Appendix C. Figure 3.7 shows two sanity checks to confirm the measurement is not impacted by these effects, for a typical measurement of the thermal conductivity of graphene using the embedded heat spreader method (Fig. 3.1). This measurement requires plotting the electrical resistance of the microfabricated line heater as a function of Joule heating power through it, and extracting its slope. In Fig. 3.7(a), we apply a positive DC current, and increase it through a sequence of 5 values from low to high. This plot is all that is strictly necessary to extract the slope and calculate k of the graphene sample. However, to help verify the raw data, as a check in Fig. 3.7(b) we then decreased

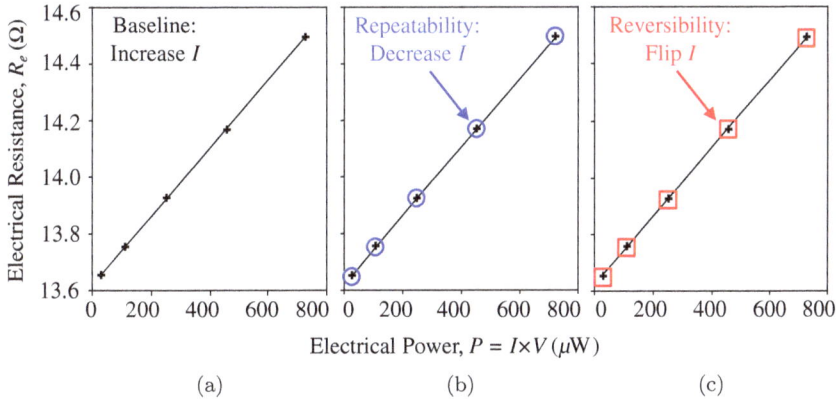

Fig. 3.7 Repeatability and reversibility checks on a DC measurement through the heater line of a graphene sample as shown in Fig. 3.1. The goal is to confirm the measurements are immune from such electrical artifacts as Seebeck voltages, low-frequency drifts, and power-line noise. (a) Baseline measurement monotonically increasing a positive DC current. (b) Repeatability and hysteresis check by then decreasing the positive DC current from high to low. (c) Reversibility check by applying a negative DC current.

the DC current through the same 5 values from high to low. The fact that the two sets of measurements (black and blue) overlap tightly with each other confirms that the waiting time between two adjacent data points is long enough for the system to reach steady state. This overlap also confirms that the system is free of significant hysteresis or drifts. Finally, in Fig. 3.7(c), we flip the polarity and apply negative DC currents, with the same magnitude but opposite sign as the 5 current values used in Fig. 3.7(a). Again, the two sets of data in Fig. 3.7(c) overlap very closely with each other, which confirms that these measurements are free from thermoelectric effects and DC offsets.

AC measurements are immune from most of the nonideal effects discussed in the preceding paragraph (see details in Appendix C). However, other artifacts are possible in principle, especially at high frequencies, such as parasitic capacitive coupling between leads, or even among microfabricated electric contact pads mediated through the dielectric isolation layer and an electrically conductive substrate. Figure 3.8 shows a scaling check to help confirm the measured signal is dominated by thermal, rather than capacitive or inductive, effects. The key is that the thermal voltages in a 3ω measurement scale as the cube of the driving current ($\Delta V_{\text{therm}} \propto I_{1\omega,\text{rms}}^3$). This scaling

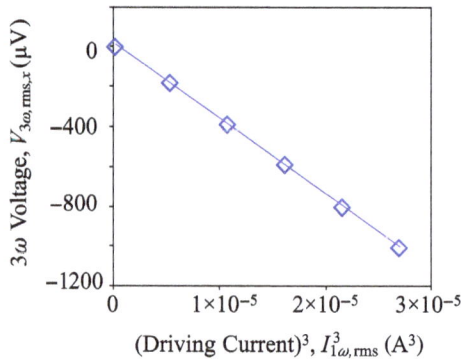

Fig. 3.8   Scaling checks on a 3ω measurement to confirm the voltage response from the heater line is thermal in origin, without significant artifacts from capacitive coupling or AC offsets. Data are obtained from one of the samples in [19].

is quite general, and applies to both the first and third harmonics, both in-phase and out-of-phase, for all 3ω measurements without a DC offset [20]. On the other hand, we would expect artifacts from capacitive or inductive coupling to follow a linear response in the electrical domain, $\Delta V_{\text{cap.coupling}} = I_{1\omega,\text{rms}} \times Z_{e,\text{parasitic}} \propto I^1_{1\omega,\text{rms}}$, where $Z_{e,\text{parasitic}}$ is some complex electrical impedance representing the parasitic coupling. Therefore, confirming that the voltage response follows a straight line when plotted against $I^3_{1\omega,\text{rms}}$ is a strong indication that the voltages indeed are thermal in origin. Thus, the scaling seen in Fig. 3.8 agrees with the prediction from Eq. (2.2), and confirms that this measurement is dominated by thermal signals rather than capacitive coupling artifacts or AC offsets.

References

[1] W. Jang, Z. Chen, W. Bao, C. N. Lau and C. Dames, "Thickness-dependent thermal conductivity of encased graphene and ultrathin graphite," *Nano Lett.* **10**(10), pp. 3909–3913, 2010.

[2] M. F. Modest, *Radiative Heat Transfer*, 2nd ed. Academic Press, 2003.

[3] K. S. Novoselov, D. Jiang, F. Schedin, T. J. Booth, V. V. Khotkevich, S. V. Morozov and A. K. Geim, "Two-dimensional atomic crystals," *Proc. Natl. Acad. Sci.* **102**(30), pp. 10451–10453, 2005.

[4] S. Lee and D. G. Cahill, "Heat transport in thin dielectric films," *J. Appl. Phys.* **81**(6), pp. 2590–2595, 1997.

[5] C. Dames, "Measuring the thermal conductivity of thin films: 3 omega and related electrothermal methods," *Annu. Rev. Heat Transf.* **16**, pp. 7–49, 2013.

[6] A. D. McConnell and K. E. Goodson, "Thermal conduction in silicon micro- and nanostructures," *Annu. Rev. Heat Transf.* **14**, pp. 129–168, 2005.

[7] D. G. Cahill, "Thermal conductivity measurement from 30 to 750 K: the 3-omega method," *Rev. Sci. Instrum.* **61**(2), pp. 802–808, 1990.

[8] D. G. Cahill and T. H. Allen, "Thermal conductivity of sputtered and evaporated SiO$_2$ and TiO$_2$ optical coatings," *Appl. Phys. Lett.* **65**(3), p. 309, 1994.

[9] E. T. Swartz and R. O. Pohl, "Thermal resistance at interfaces," *Appl. Phys. Lett.* **51**, p. 2200, 1987.

[10] Y. S. Touloukian, *Thermophysical Properties of Matter*. New York: IFI/Plenum.

[11] D. Li, Y. Wu, P. Kim, L. Shi, P. Yang and A. Majumdar, "Thermal conductivity of individual silicon nanowires," *Appl. Phys. Lett.* **83**(14), pp. 2934–2936, 2003.

[12] Z. Wang, J. E. Alaniz, W. Jang, J. E. Garay and C. Dames, "Thermal conductivity of nanocrystalline silicon: Importance of grain size and frequency-dependent mean free paths," *Nano Lett.* **11**(6), pp. 2206–2213, 2011.

[13] C. Dames, "Microscale conduction," in *Heat Conduction*, 3rd ed., Lead Author Latif Jiji, Springer, 2009.

[14] C. Dames and G. Chen, "Thermal conductivity of nanostructured thermoelectric materials," in *Thermoelectrics Handbook: Macro to Nano*, Taylor & Francis, 2006.

[15] F. Yang and C. Dames, "Mean free path spectra as a tool to understand thermal conductivity in bulk and nanostructures," *Phys. Rev. B* **87**(3), p. 35437, 2013.

[16] N. W. Ashcroft and N. David Mermin, *Solid State Physics.* Brooks/Cole: Thomson Learning, Inc., 1976.

[17] J. M. Ziman, *Electrons and Phonons.* New York: Oxford University Press, 1960.

[18] E. S. Toberer, L. L. Baranowski and C. Dames, "Advances in thermal conductivity," *Annu. Rev. Mater. Res.* **42**(1), pp. 179–209, 2012.

[19] Z. Chen, W. Jang, W. Bao, C. N. Lau and C. Dames, "Thermal contact resistance between graphene and silicon dioxide," *Appl. Phys. Lett.* **95**(16), pp. 161910–161913, 2009.

[20] C. Dames and G. Chen, "1ω, 2ω, and 3ω methods for measurements of thermal properties," *Rev. Sci. Instrum.* **76**(12), p. 124902, 2005.

Chapter 4

Uncertainty and Sensitivity Analysis

"Offered the choice between mastery of a 5-foot shelf of analytical statistics books and middling ability at performing statistical Monte Carlo simulations, we would surely choose to have the latter skill."

— W. H. Press *et al.*, Numerical Recipes

4.1. Introduction

In this chapter, we highlight the importance of uncertainty and sensitivity analysis: not only in assigning upper and lower bounds to the measurement result, but also for identifying which possible changes to an experiment can best enhance its accuracy. We first discuss the conventional partial derivative (PD) method of uncertainty analysis and its limitations. Next we introduce the concept of sensitivity, which is a powerful tool to identify the dominant uncertainty sources. Last we introduce a Monte Carlo (MC) scheme, and discuss in detail a recipe for its implementation. The examples in this chapter will often refer back to the heat spreader method which has been discussed in Chapter 1 (e.g., Fig. 1.8) and Chapter 2 (e.g., Fig. 2.13).

While we expect many readers have some acquaintance with the PD method, this chapter may be their first introduction to the MC approach which therefore will be explained more thoroughly. The MC scheme is particularly powerful when some of the parameters have non-Gaussian probability distributions, which violates one of the key assumptions of the PD method. For example, for a parameter with large uncertainty it often makes more sense to assume it follows a lognormal rather than normal probability distribution, thereby ensuring its value can never be negative.

4.2.   Preliminaries

As shown in Fig. 4.1, we may think of a thermal measurement as a
process to determine one or more properties (**a**) of a sample based on
control variables (**X**) and response variables (**Y**). In general, **a**, **X**,
and **Y** are all vectors, which could be composed of many components.
For example, in the heat spreader method (e.g., Figs. 1.8 and 2.13),
the components of the **X** vector include the Joule heating power, all
the geometries, and the thermal properties of the various constituents
except the unknown thermal conductivity of the graphene flake; the
Y vector has three components, namely the temperatures recorded
by the three sensors; and **a** is the thermal conductivity of the
graphene flake, a scalar. Another example is the calibration of the
temperature-dependent electrical resistance of a heater line, to be
discussed in Sec. 4.4.2. In this case, **X** and **Y** are the N calibration
temperatures and corresponding electrical resistance values, both of
which are vectors of length N, and $\mathbf{a} = \{r_0, \theta, \Delta\}$ is a vector of three
model parameters.

The primary goal of a thermal measurement is to obtain one
or more numbers, **a**. For example, the thermal conductivity of an
undoped silicon wafer at room temperature was measured to be
$k = 148$ W/m $\cdot$ K. However, there is more information we should
extract. First, since no measurement can be completely free of errors
and uncertainties, we want to know how close our measured number
is likely to be to the truth. Second, if this accuracy is deemed
insufficient, we may also want to know the best ways to further

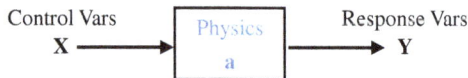

Fig. 4.1   A generic thermal measurement can be thought of as an algorithm to
extract the thermal properties (the physics variables, **a**) of a material based on
various control variables and response variables. Here we group all the control
variables as an abstract vector **X**, and likewise **a** and **Y**, with examples and
details in the main text.

reduce the uncertainty of our measurement. Correspondingly, this chapter tries to address two key questions:

- *How confidently can we claim that the true value is within a specific range,* $[a_{\min}, a_{\max}]$? For example, using the 3ω method, we may conclude that we are 95% confident that the true thermal conductivity of the undoped silicon wafer at room temperature is somewhere in the range [128, 168] W/m · K.
- *Which control or response variables dominate the total uncertainty?* For example, using the 3ω method to measure the thermal conductivity k of silicon requires knowledge of the length l, the width $2b$, the electrical resistivity ρ, and the temperature coefficient α, of the heater line. For each of these we must already have an estimate of their true value and uncertainty: $l, u_l, b, u_b, \rho, u_\rho, \alpha, u_\alpha$. But some of these input uncertainties may be much more important for the ultimate uncertainty in k than others. For example, we may find that 75% of the total uncertainty in k can be traced back to the uncertainty in α, while only 5% of the uncertainty in k can be attributed to the uncertainty in l. In this case, if we need to further tighten the confidence interval (CI) on k, our efforts would be much better focused on reducing u_α than u_l.

We have at least two approaches to answer the questions above. The most straightforward approach, as depicted in Fig. 4.2, is to conduct multiple measurements and analyze the results statistically. Continuing with the heat spreader method as an example, multiple measurements include repeating measurements on a single sample, or much better preparing nominally identical samples and measuring each of them. However, in many practical cases we are constrained by the time and expense of preparing samples and running the experiment, such that we often may only have only one sample of any given type. In this scenario, if we know the uncertainties of the control and response variables there is still a second approach to answering the two questions, by the propagation of errors (Fig. 4.3).

While the first approach (Fig. 4.2) is straightforward, in this chapter we focus on the second approach (Fig. 4.3), the propagation

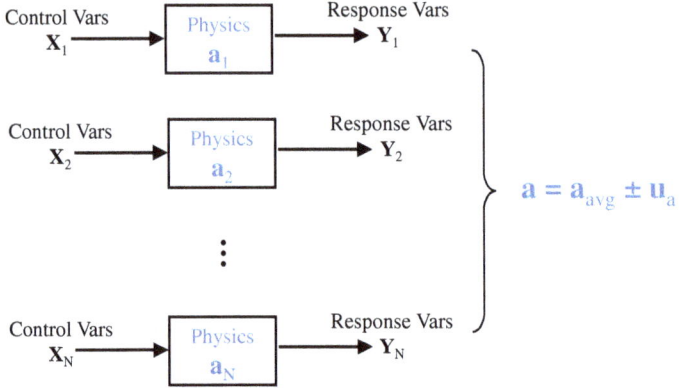

Fig. 4.2   An approach to uncertainty estimation that is straightforward in principle but sometimes impractical: multiple measurements. For example, in the heat spreader experiments, we may prepare nominally identical samples and measure each of them, or repeat measurements on a single sample. In each of the N measurements we extract a best estimate for the physics $\mathbf{a}$, which are then analyzed to report our best estimate $\mathbf{a}_{avg}$ and its uncertainty $\mathbf{u}_a$.

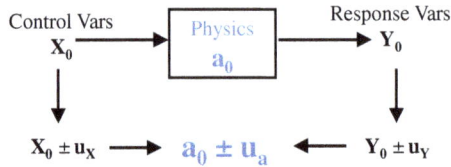

Fig. 4.3   The uncertainty estimation approach focused on in this chapter: a single measurement (subscript 0) combined with error propagation. We will discuss both PD and MC methods for error propagation.

of errors. To this end we will introduce two methods, a partial derivative method and a Monte Carlo method, each with its own pros and cons. We shall now discuss these two methods with detailed examples.

4.3.   Partial derivative method

The partial derivative (PD) method is a conventional method for uncertainty analysis (see [7, p. 3]). Although it is subject to certain restrictions listed at the end of this section, the key merit of the

PD method is that it cannot only give a CI, but also indicate the dominant error sources. It also has a clear and appealing analytical formulation.

We first present a general framework for the PD method. Because the control and response variables are treated equivalently here, they can be concatenated as a new vector $\mathbf{Z} = [\mathbf{X}; \mathbf{Y}]$. We then express the property vector $\mathbf{a}$ as a function of $\mathbf{Z}$, namely,

$$\mathbf{a} = f(\mathbf{Z}). \tag{4.1}$$

Also, for simplicity in notation we analyze only one component of $\mathbf{a}$, and write it as a scalar, $a = f(z_1, z_2, \ldots, z_n)$. The results are readily generalized to the other components of $\mathbf{a}$.

The essence of the PD method is that the uncertainty of a can be expressed as

$$u_a = \sqrt{\left(\frac{\partial a}{\partial z_1} u_{z_1}\right)^2 + \left(\frac{\partial a}{\partial z_2} u_{z_2}\right)^2 + \cdots + \left(\frac{\partial a}{\partial z_n} u_{z_n}\right)^2}, \tag{4.2}$$

where u_{z_i} is the uncertainty in the ith variable z_i.

To obtain a relative uncertainty, we rewrite Eq. (4.2) as

$$\frac{u_a}{a} = \sqrt{\sum_{i=1}^{n} \left(S_i^a \times \frac{u_{z_i}}{z_i}\right)^2}, \tag{4.3}$$

which defines a dimensionless *sensitivity* [1, 8, 9]

$$S_{z_i}^a = \frac{z_i}{a} \frac{\partial a}{\partial z_i} = \frac{\partial(\ln a)}{\partial(\ln z_i)}. \tag{4.4}$$

For example, $S_{z_i}^a = -10$ means that a 1% increase in z_i will cause a 10% decrease in a.

Based on $S_{z_i}^a$, we also define an uncertainty contribution,

$$c_{z_i}^a = |S_{z_i}^a| \times (u_{z_i}/z_i), \tag{4.5}$$

for each variable, such that

$$\frac{u_a}{a} = \sqrt{\sum_{i=1}^{n} (c_{z_i}^a)^2}. \tag{4.6}$$

We now demonstrate this framework with an example, which analyzes a 12-layer-thick graphene sample [1] using the heat spreader method (Secs. 1.7 and 2.2.2(F)). In this example, a is the thermal conductivity of graphene, $k_{\text{gr},\|}$. As summarized in Table 4.1, the control variables ($\mathbf{X}$) are the geometries and the thermal properties of the sample stack, and the response variables ($\mathbf{Y}$) are the temperature response recorded by the metal-line sensors. Note that in this sample only the first two sensors were working [1].

Table 4.1 outlines the relative uncertainty (u_{z_i}/z_i) of each input parameter corresponding to one standard deviation of a Gaussian distribution. The two largest relative uncertainties are in the thermal contact resistance between graphene and SiO_2 and the thermal conductivity of the substrate, with $u_{R''_{c,\text{gr-ox}}}/R''_{c,\text{gr-ox}} = 50\%$ and $u_{k_{\text{sub}}}/k_{\text{sub}} = 20\%$, respectively.

To calculate the sensitivity ($S_{z_i}^{k_{\text{gr},\|}}$), we numerically evaluate the PDs (Eq. (4.4)) using small perturbations of each parameter around its typical value. These calculations show that this k_{gr} is most *sensitive* to the following four parameters: (a) distance between the heater and the first sensor with $S_{d_{\text{HtrToS1}}}^{k_{\text{gr},\|}} = 6.41$, (b) the thickness and (c) the thermal conductivity of the bottom oxide layer with $S_{t_{\text{BotOx}}}^{k_{\text{gr},\|}} = -3.84$ and $S_{k_{\text{BotOx}}}^{k_{\text{gr},\|}} = 3.54$, respectively, and (d) the temperature response of sensor #1 with $S_{\Delta T_{\text{S1}}}^{k_{\text{gr},\|}} = 3.06$.

With u_{z_i}/z_i and $S_{z_i}^{k_{\text{gr},\|}}$ above, we obtain the uncertainty contribution ($c_{z_i}^{k_{\text{gr},\|}}$) for each parameter using Eq. (4.5). The three most important *contributions* come from $k_{\text{BotOx}}, k_{\text{sub}}$, and ΔT_{S1}, with $c_{k_{\text{BotOx}}}^{k_{\text{gr},\|}} = 17.7\%, c_{k_{\text{sub}}}^{k_{\text{gr},\|}} = 15.6\%$, and $c_{\Delta T_{\text{S1}}}^{k_{\text{gr},\|}} = 15.3\%$.

Proceeding similarly with all other parameters of Table 4.1, we finally combine the results using Eq. (4.6) to estimate the 68% CI of this 12-layer-thick graphene sample as [64, 119] W/m · K, and the 95% CI as [37, 146] W/m · K.

We close this section by noting some limitations of the PD method. There are three key assumptions of this framework: the uncertainties in all control and responses variables are uncorrelated, the errors are small perturbations, and the errors follow Gaussian distributions [7]. In many cases, the latter two assumptions may be

Table 4.1 Example of uncertainty analysis in the heat spreader method using the PD method, for a 12-layer-thick graphene sample [1].

Input parameters		Typical value, z_i	Units	Relative uncertainty, u_{z_i}/z_i	Sensitivity, $S_{z_i}^{k_{\mathrm{gr},\parallel}} \equiv \dfrac{\partial(\ln k_{\mathrm{gr},\parallel})}{\partial(\ln z_i)}$	Contribution, $c_{z_i}^{k_{\mathrm{gr},\parallel}} \equiv |S_{z_i}^{k_{\mathrm{gr},\parallel}}| \times (u_{z_i}/z_i)$	Comment on uncertainty estimate
Control variables, X							
Thickness of top and bottom oxide layers	t_{TopOx}	0.025	μm	12.0%	−0.10	1.2%	Crystal monitor and stylus profilometer
	t_{BotOx}	0.32		1.6%	−3.84	6.0%	Ellipsometer and SEM
Center-to-center distances between metal lines	d_{HtrToS1}	0.87		0.3%	6.41	2.2%	Placement accuracy of e-beam lithography patterning; measurements using SEM
	d_{HtrToS2}	1.61		0.2%	−0.80	0.2%	
Width of heater	$2b$	0.50		2.0%	−0.44	0.4%	
Width of T sensors	$2b_{\mathrm{sens}}$	0.24		2.1%	−0.20	0.9%	
Thermal conductivities of oxide layers, Si substrate, and gold electrodes	k_{TopOx}	0.95	W/m · K	5.0%	−0.11	0.6%	Comparison between 3ω measurements, FEM fitting of control samples, and lithography values [2, 3]
	k_{BotOx}	1.43		5.0%	3.54	17.7%	
	k_{sub}	93		20%	0.78	15.6%	
	$k_{\mathrm{electrode}}$	141		9.9%	0.06	0.6%	Measured resistivity + Wiedemann–Franz law
TBR between gr and ox	$R''_{\mathrm{c,gr-Ox}}$	9.5	10^{-9} W/m² · K	50%	−0.12	6.0%	Measured by 3ω [4]
Response variables, Y							
Sensor response: T rise per unit heating power	ΔT_{S1}	8.35	10^{3} K/W	5.0%	3.06	15.3%	Consequence of measurements and fitting with Bloch–Grüneisen formula [6]
	ΔT_{S2}	2.70		5.0%	0.36	1.8%	
Combine X and Y as a new vector $\mathbf{Z} = [\mathbf{X}; \mathbf{Y}]$, of length $N = 13$							

The control variables $\mathbf{X}$ and the response variables $\mathbf{Y}$ are grouped as new variables $\mathbf{Z} = [\mathbf{X}; \mathbf{Y}]$. All uncertainties correspond to the standard deviations of Gaussian distributions, except k_{sub} and $R''_{\mathrm{c,gr-Ox}}$ which are taken to follow lognormal distributions. Two important parameters for assessing this measurement are the sensitivities $S_{z_i}^{k_{\mathrm{gr},\parallel}}$ (Eq. (4.4)), which indicate which parameters the thermal conductivity of graphene is most sensitive to, and the uncertainty contribution $c_{z_i}^{k_{\mathrm{gr},\parallel}}$ (Eq. (4.5)), which reveals the dominant error contributors to be $c_{k_{\mathrm{BotOx}}}^{k_{\mathrm{gr},\parallel}} = 17.7\%$, $c_{k_{\mathrm{sub}}}^{k_{\mathrm{gr},\parallel}} = 15.6\%$, and $c_{\Delta T_{S1}}^{k_{\mathrm{gr},\parallel}} = 15.3\%$.

violated, thus limiting the method's applicability. In such cases, the Monte Carlo method is a good complement, which we turn to next.

4.4. Monte Carlo method

The Monte Carlo (MC) scheme (see [5, p. 807]) is a good complement to the PD method, especially when one or more of the key assumptions of the PD method are violated. For example, when the uncertainties of the input parameters cannot be treated as small perturbations, and even do not follow Gaussian distributions (Sec. 4.4.1) [1], the MC method is a better choice for uncertainty analysis. In addition, the MC method is particularly powerful for evaluating the uncertainty of a nonlinear fit (Sec. 4.4.2) [4].

4.4.1. *Detailed walkthrough of the MC method: An example using variables with large and non-Gaussian uncertainty distributions about their mean value*

One example in this category is the k_{gr} analysis shown in Table 4.1, in which the contact resistance between graphene and SiO_2 has rather large uncertainty, violating one of the basic assumptions of the PD uncertainty analysis described in Sec. 4.3.

To begin the MC scheme, we must first identify the control variables ($\mathbf{X}$), the response variables ($\mathbf{Y}$), the physical model, and the fitting parameters ($\mathbf{a}$). As above, $\mathbf{X}$ is a vector composed of all the geometries and the thermal properties input to the FEM model (see Fig. 2.13 and Table 4.1), except k_{gr}; $\mathbf{Y}$ is a vector composed of the temperatures measured by the three resistive thermometers; the physical model used for fitting is the three-dimensional (3D) FEM model; and $\mathbf{a}$ is $k_{gr,\parallel}$, a scalar.

We explain the detailed steps of the MC analysis recipe through the following extended example, numbered to correspond to the flowchart of Fig. 4.4:

(1) Obtain the best fit for the model physics vector $\mathbf{a_0}$ (a scalar $k_{gr,0}$ in this example) based on the measured control variables ($\mathbf{X}_{msrd}$),

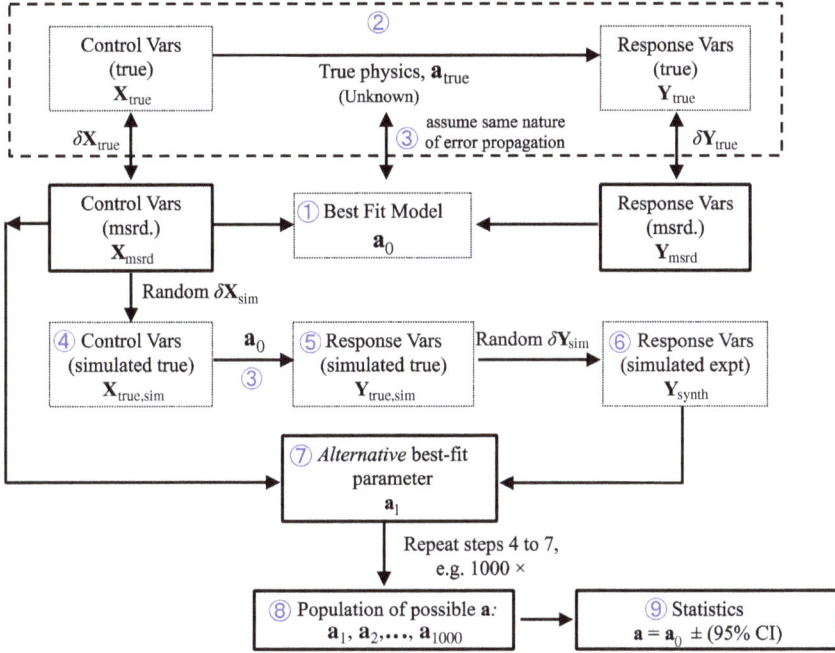

Fig. 4.4 Flowchart of an MC scheme to analyze uncertainty. The numbers in blue circles correspond to the steps described in the recipe in text.

the measured response variables ($\mathbf{Y}_{\mathrm{msrd}}$), and the physical model (the 3D FEM model). For the 12-layer-thick graphene example as discussed in Table 4.1 in Sec 4.3, the best fit was already found to be $k_{\mathrm{gr},\parallel,0} = 91.5$ W/m $\cdot$ K.

(2) Consider the relationships between the measured and true control variables ($\mathbf{X}_{\mathrm{msrd}} \leftrightarrow \mathbf{X}_{\mathrm{true}}$), the response variables ($\mathbf{Y}_{\mathrm{msrd}} \leftrightarrow \mathbf{Y}_{\mathrm{true}}$), and physics ($\mathbf{a}_{\mathrm{msrd}} \leftrightarrow \mathbf{a}_{\mathrm{true}}$, or $k_{\mathrm{gr,msrd}} \leftrightarrow k_{\mathrm{gr,true}}$ in this example). Although these true values are unknowable, we do know that the measured variables ($\mathbf{X}_{\mathrm{msrd}}$ and $\mathbf{Y}_{\mathrm{msrd}}$) are perturbed around the true variables ($\mathbf{X}_{\mathrm{true}}$ and $\mathbf{Y}_{\mathrm{true}}$) by some perturbations ($\delta\mathbf{X}_{\mathrm{true}}$ and $\delta\mathbf{Y}_{\mathrm{true}}$; here δ denotes perturbation), which are unknown but for which we must have estimates of the underlying uncertainty distributions ($u_{\mathbf{X}}$ and $u_{\mathbf{Y}}$).

Continuing with the 12-layer-thick graphene example, let us consider one entry in the $\mathbf{X}$ vector, the thermal conductivity k_{sub} of the doped silicon substrate. We can never know its true value $k_{\text{sub,true}}$, so we settle for working with our best estimate based on a separate 3ω measurement, which gave $k_{\text{sub,msrd}} = 93$ W/m · K. Further, based on our experience and experimental judgment we believe its (unknown) perturbation δk_{sub} to be drawn from a lognormal uncertainty distribution of standard deviation $u_{k_{\text{sub}}} = 19$ W/m · K (see details in Appendix H).

(3) Now comes the key assumption of the MC approach [5, p. 807]: that *the nature of error propagation in the true system* ($\mathbf{a}_{\text{true}}$) *can be well approximated by the nature of error propagation in our best-estimated understanding of the system* ($\mathbf{a}_0$). This is *not* the same as asserting $\mathbf{a}_{\text{true}} = \mathbf{a}_0$; rather, we are assuming that $\delta\mathbf{a}$ is a weak function of $\mathbf{a}$ in the vicinity of $\mathbf{a}_{\text{true}}$, so that $\delta\mathbf{a}$ can be approximated using calculations based on $\mathbf{a}_0$. So as applied to the present example, to estimate the impact of δk_{sub} on $\delta k_{\text{gr},\|}$ we use an FEM model based on $k_{\text{gr},0}$ and assume that this gives a similar result for $\delta k_{\text{gr},\|}$ as the (impossible) exercise of using an FEM model based on $k_{\text{gr},\|,\text{true}}$. This key idea is implemented using numerous "synthetic experiments", as explained next in steps (4)–(7).

(4) First, simulate a randomized $\mathbf{X}_{\text{true,sim}}$ vector representing a fair estimate of what the underlying control variables might actually have been in the real experiment, based on our beliefs about the uncertainties in $\mathbf{X}$. Each entry of $\mathbf{X}$ has its own uncertainty distribution. Importantly, these distributions may be large and/or non-Gaussian, traits which cannot be handled by the PD approach. To generate this $\mathbf{X}_{\text{true,sim}}$, we use a random number generator (hence this scheme's Monte Carlo name) to estimate a random perturbation vector $\delta\mathbf{X}_{\text{sim}}$. Thus, we have a simulated set of control variables:

$$\mathbf{X}_{\text{true,sim}} = \mathbf{X}_{\text{msrd}} - \delta\mathbf{X}_{\text{sim}}. \tag{4.7}$$

Continuing the example for the specific entry of $\mathbf{X}$ which is k_{sub}, we draw a random uncertainty perturbation δk_{sub} from the aforementioned uncertainty distribution (lognormal distribution with standard deviation $u_{k_{\text{sub}}} = 19$ W/m · K), and let us suppose for example this returned the value $\delta k_{\text{sub}} = 13$ W/m · K. Then, we have a simulated "true" value $k_{\text{sub,true,sim}} = 93 - 13 = 80$ W/m · K. This is similarly applied to all the other entries of $\mathbf{X}$ to generate an $\mathbf{X}_{\text{true,sim}}$ vector.

(5) Next, based on these simulated true control variables ($\mathbf{X}_{\text{true,sim}}$) from step (4) and the best fit $k_{\text{gr},0}$ from step (1), we generate the simulated true response variables, $\mathbf{Y}_{\text{true,sim}}$:

$$\mathbf{X}_{\text{true,sim}} \xrightarrow[\text{Best Fit Model}]{\mathbf{a}_0} \mathbf{Y}_{\text{true,sim}}. \tag{4.8}$$

In terms of the k_{sub} example, we revisit the FEM model and re-run it using $k_{\text{sub,true,sim}} = 80$ W/m · K (and likewise all the other simulated perturbed $\mathbf{X}_{\text{true,sim}}$ values), but always using the original best estimate value $k_{\text{gr},\parallel,0} = 91.5$ W/m · K. This FEM run will yield a specific set of response variables $\mathbf{Y}_{\text{true,sim}}$, in this example the sensor temperature rises per unit heating power, $\Delta T_{\text{S1}}, \Delta T_{\text{S2}}$, and ΔT_{S3}. Let us focus on just one of these three, and suppose that this particular run gives $\Delta T_{\text{S1,true,sim}} = 9000$ K/W.

(6) We must also account for the fact that our measurements could not have accessed those $\mathbf{Y}_{\text{true,sim}}$ directly, so they must be further randomized to account for their experimental uncertainty. This again means using a random number generator to estimate a random perturbation vector $\delta\mathbf{Y}_{\text{sim}}$, based on our understanding of the probability distributions of uncertainty for each member of $\mathbf{Y}$, distributions which again can be large and/or non-Gaussian. Thus we will have a synthetic set of experimental response variables $\mathbf{Y}_{\text{synth}}$ corresponding to this particular $\mathbf{Y}_{\text{true,sim}}$,

$$\mathbf{Y}_{\text{synth}} = \mathbf{Y}_{\text{true,sim}} + \delta\mathbf{Y}_{\text{sim}}. \tag{4.9}$$

Continuing with the discussion of the ΔT_{S1} entry of $\mathbf{Y}_{\text{true,sim}}$, we believe our thermometry to have $u_{\Delta T_{\text{S1}}} = \pm 5\%$ uncertainty. Supposing a call of the random number generator gives a value $\delta(\Delta T_{\text{S1}}) = +2\%$, this corresponds to a synthetic experiment with a "measured" temperature rise $\Delta T_{\text{S1,synth}} = 9000 + 180 = 9180$ K/W.

(7) Now, obtain a best fit $k_{\text{gr},1}$ based on the **measured** control variables ($\mathbf{X}_{\text{msrd}}$) and the **synthesized** response variables ($\mathbf{Y}_{\text{synth}}$).

(8) Repeat steps (4)–(7) numerous (for example, $N \approx 1000$) times, generating new random perturbations $\delta\mathbf{X}_{\text{sim}}$ and $\delta\mathbf{Y}_{\text{sim}}$ every time,

to obtain a population $\{k_{\mathrm{gr},\parallel,1}, k_{\mathrm{gr},\parallel,2}, \dots, k_{\mathrm{gr},\parallel,N}\}$. If the physical model being fit for had more than one parameter, this would instead be a population of vectors, $\{\mathbf{a}_1, \mathbf{a}_2, \dots, \mathbf{a}_N\}$.

(9) Now it is straightforward to analyze the statistics of this population of synthetic experiments $\{k_{\mathrm{gr},\parallel,1}, k_{\mathrm{gr},\parallel,2}, \dots, k_{\mathrm{gr},\parallel,N}\}$ to obtain our estimated uncertainty for $k_{\mathrm{gr},\parallel,0}$ with a specified CI. This invokes the key assumption stated at the beginning of step (3) that the spread of the synthetic population is a good estimate for the spread of an (unknowable) population of many true experiments. To obtain this CI:

- First, sort the simulated $\{k_{\mathrm{gr},\parallel,1}, k_{\mathrm{gr},\parallel,2}, \dots, k_{\mathrm{gr},\parallel,N}\}$ in ascending order.
- Next, to evaluate a CI of $(1 - \chi) \times 100\%$, we throw out the first and last $[(\chi/2) \times N - 1]$ points of the sorted list. For example, the commonly reported 95% CI corresponds to $\chi = 0.05$. So if we have generated $N = 1000$ synthetic experimental points, this means throwing out the 24 largest and 24 smallest values of the sorted $\{k_{\mathrm{gr},1}, k_{\mathrm{gr},2}, \dots, k_{\mathrm{gr},1000}\}$ list.
- Finally, the smallest and the largest remaining data give the lower and the upper bound of the $(1 - \chi) \times 100\%$ CI.

Applying this recipe to the 12-layer-thick graphene example, we found the 95% CI as [41.9, 190.1] W/m · K [1]. Note here we were seriously limited by the computational time of the 3D FEM model fitting, so that we used a much smaller $N = 80$. Note also that the MC analysis does not in any way change the best estimate value $k_{\mathrm{gr},\parallel,0} = 91.5$ W/m · K, which as depicted in step (1) of the flowchart used only the directly measured $\mathbf{X}$ and $\mathbf{Y}$.

We end by comparing the results from the PD and MC frameworks (Fig. 4.5). While the PD framework results in a 95% CI of [37, 146] W/m · K, the MC framework gives a 95% CI of [42, 190] W/m · K. Note that in both the PD and the MC frameworks, the 95% CI is rather large, calling into question whether the PD method is even appropriate for this analysis; while the MC method is applicable even for large CI ranges.

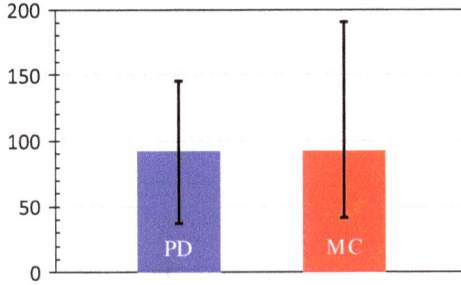

Fig. 4.5 Comparison between the PD and MC results on the 12-layer-thick graphene sample [1]. The nominal value is 91.5 W/m · K for both methods, obtained from the best fit to the experimental measurement. The PD method has a 95% CI of [37, 146] W/m · K, while the MC method has a 95% CI of [42, 190] W/m · K. Due to the broad k_{gr} CI, as well as the large and non-Gaussian uncertainty distributions of the inputs k_{sub} and $R''_{c,gr-ox}$, one would question whether the PD method is appropriate for this analysis.

4.4.2. *Another application of the MC method: Quantifying the uncertainty of a nonlinear fit*

Another category of problems that are especially suitable for an MC method is the quantification of the uncertainty of a nonlinear fit which involves a complicated physical model. A good example is again motivated by the analysis in Table 4.1, which indicates that further reducing the uncertainty of the temperature rise of sensor S1 is one of the best ways to improve the measurement accuracy of k_{gr}, since both its sensitivity $S_{\Delta T_{S1}}^{k_{gr,\parallel}} = 3.06$ and uncertainty contribution $c_{\Delta T_{S1}}^{k_{gr,\parallel}} = 15.3\%$ are among the highest in the table. The underlying calibration is the relation between the measured sensor resistance ($R_{e,mrsd}$) and its corresponding temperature (T_{set}). The key quantity to be calibrated is the slope $m = dR_{e,mrsd}/dT_{set}$ with its corresponding uncertainty u_m, where both m and u_m could be functions of T_{set}.

Table 4.2 shows a typical calibration dataset of $R_{e,mrsd}$ (second column) over a wide temperature range of T_{set} from 10 K to 310 K (first column). Before applying the MC scheme, we first review a simpler least-squares analysis, which uses a textbook analytical

Table 4.2 Comparison of two approaches for analyzing a resistance thermometry calibration dataset, including uncertainty (see also Fig. 4.6 for the corresponding visualization).

Temperature, T_{set} [K]	Resistance, $R_{e,mrsd}$ [Ω]	Textbook approach & linear fit of the slope (80K ≤ T_{set} ≤310K)		Monte Carlo approach & Bloch-Grüneisen fit of the slope (10K ≤ T_{set} ≤310K)			
		m_{linear} [10^{-5} Ω/K]	$u_{m_{B-G}}$ [10^{-5} Ω/K]	Best fit to experiments, $m_{B-G,0}$ [10^{-5} Ω/K]	Lower bound, $m_{B-G,LB}$ [10^{-5} Ω/K]	Upper bound, $m_{B-G,UB}$ [10^{-5} Ω/K]	$u_{m_{B-G}} = \left(\dfrac{m_{B-G,UB} - m_{B-G,LB}}{2}\right)$ [10^{-5} Ω/K]
310	3.38			826	821	830	4.7
250	2.87			832	827	836	4.6
216	2.59			837	8.33	842	4.5
188	2.35			844	839	849	4.9
164	2.15			852	846	858	5.8
142	1.96			863	857	869	6.2
124	1.80			876	869	882	6.8
106	1.65			893	885	900	7.4
92	1.52	854	7.3	910	902	917	7.7
80	1.41			928	919	935	7.7
71	1.32			941	933	948	7.1
63	1.25			951	944	956	5.9
55	1.17			953	947	959	6.2
48	1.11			940	928	954	13.0
42	1.06			907	885	930	22.6
37	1.01			852	819	888	34.5
33	0.98			781	738	828	44.8
29	0.95			680	628	737	54.3
22	0.91			424	372	487	57.5
17	0.89			216	181	262	40.7
13	0.88			87	70	110	20.0
10	0.88			32	25	41	7.8

The first two columns give the raw measurement data. The next two columns apply a linear $R_c(T)$ model which is amenable to a well-known textbook calculation of the 95% CI, though this model is only suitable for the higher temperature data. The remainder of the table presents an MC approach to analyze a nonlinear Bloch-Grüneisen model fit to the complete dataset. See text for details.

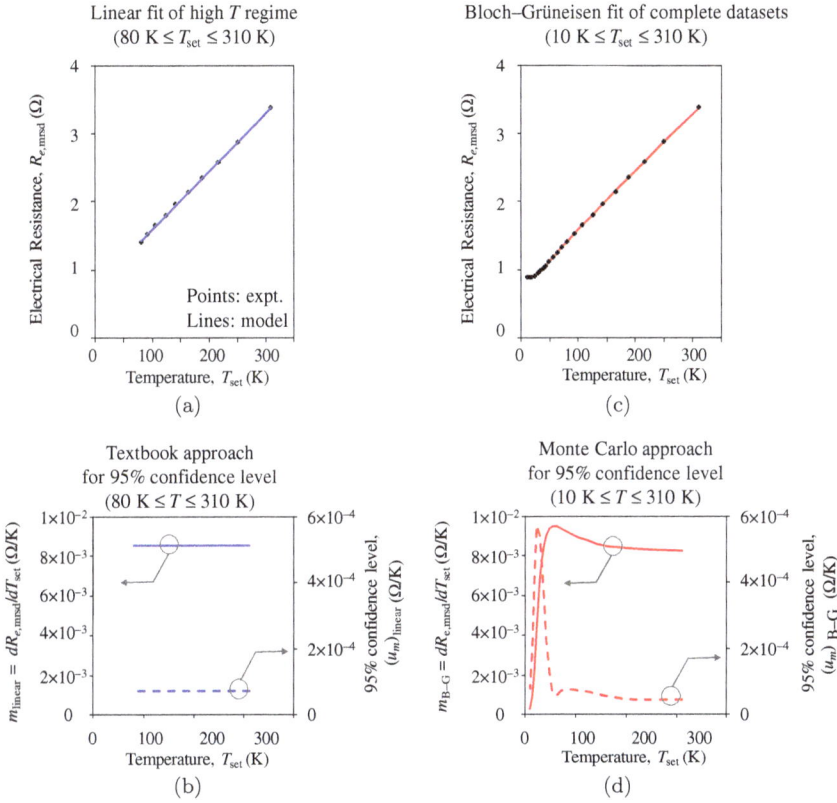

Fig. 4.6 A textbook approach to analyze the 95% CI of a linear fit to the high temperature dataset of $(T_{set}, R_{e,mrsd})$ in Table 4.2 vs. an MC approach to analyze the 95% CI of a nonlinear fit to the complete temperature dataset. (a) High temperature dataset (points) and the linear fitting (line). (b) The slope (solid line; left axis) of the linear fit in panel (a), and its corresponding 95% CI (dashed line; right axis). (c) Complete dataset (points) and the nonlinear fitting (line). (d) The slope (solid line; left axis) of the nonlinear fit in panel (c), and its corresponding 95% CI (dashed line; right axis).

expression to quantify the uncertainty of a linear fit to the high temperature regime ($T_{set} \geq 80$ K). This side calculation helps us appreciate how complicated the analytical expression is even for such a simple scenario.

Figure 4.6(a) isolates the higher-temperature regime ($T_{set} \geq 80$ K) in which a linear fit is generally taken to be adequate. Following a standard least-squares approach (see, [11, p. 377]), the slope of a linear fit, $m_{\text{linear}} = dR_{e,\text{High-}T}/dT$, is

$$m_{\text{linear}} = \frac{S_{xy}}{S_{xx}}, \tag{4.10}$$

where

$$S_{xx} = \sum_{i=1}^{n} T_{\text{set},i}^2 - \left(\sum_{i=1}^{n} T_{\text{set},i}\right)^2 \Bigg/ n, \tag{4.11}$$

$$S_{xy} = \sum_{i=1}^{n} (T_{\text{set},i} \cdot R_{e,\text{mrsd},i}) - \left(\sum_{i=1}^{n} T_{\text{set},i}\right) \cdot \left(\sum_{i=1}^{n} R_{e,\text{mrsd},i}\right) \Bigg/ n, \tag{4.12}$$

n is the number of data points, and $R_{e,\text{mrsd},i}$ is the electrical resistance measured at a specific temperature $T_{\text{set},i}$. Note that this m_{linear} is by definition independent of T.

The uncertainty in this least-squares m_{linear} is also well known (see [11, p. 390]). Although this approach does not explicitly take into account the uncertainty distributions of the raw measured temperature ($\mathbf{u}_T$) and resistance ($\mathbf{u}_R$) values, their impact is still felt through the random scatter in the $(T_{\text{set}}, R_{e,\text{mrsd}})$ points. At the $(1 - \chi) \times 100\%$ CI, the result is

$$u_{m_{\text{linear}}} = t_{\chi/2,n-2}\sqrt{\frac{V}{S_{xx}}}, \tag{4.13}$$

where the unbiased estimator of the variance is (see [11, p. 379])

$$V = \frac{\sum_{i=1}^{n} (R_{e,\text{mrsd},i} - R_{e,\text{fit},i})^2}{n - 2}$$

$$= \frac{[\sum_{i=1}^{n} R_{e,\text{mrsd},i}^2 - (\sum_{i=1}^{n} R_{e,\text{mrsd},i})^2/n] - m_{\text{linear}} \cdot S_{xy}}{n - 2}, \tag{4.14}$$

and $t_{\alpha/2,n-2}$ is the Student's t-statistic for the upper $(\chi/2) \times 100\%$ confidence threshold in this distribution with $(n - 2)$ degrees of freedom. For example, for $\chi = 0.05$, the upper 2.5% point of the t distribution with $n = 10$ data points of $(T_{\text{set}}, R_{e,\text{mrsd}})$ is determined from standard tables to be $t_{(0.05/2),(10-2)} = 2.306$.

Applying Eqs. (4.10)–(4.14) to the first 10 data points of $(T_{\text{set}}, R_{\text{mrsd}})$, we obtain $m_{\text{linear}} = 854 \times 10^{-5}\,\Omega/\text{K}$ and $u_{m_{\text{linear}}} = 7.3 \times 10^{-5}\,\Omega/\text{K}$, as shown in the third and fourth columns of Table 4.2 and the solid and dashed lines in Fig. 4.6(b). This yields a 95% CI of $[(m_{\text{linear}} - u_{m_{\text{linear}}}), (m_{\text{linear}} + u_{m_{\text{linear}}})]$ for the temperature range 80–310 K.

Due to the flattening in the $R_e(T)$ function at low T (Table 4.2; Fig. 4.6(c)), fitting the full T range clearly requires some kind of nonlinear model function. Rather than following a purely empirical approach such as higher order polynomials, we use the physically-based Bloch–Grüneisen formula [4, 6]:

$$R_{e,\text{B–G}}(T) = r_0 + 4\Delta \left(\frac{T}{\theta}\right)^5 \int_0^{\theta/T} \frac{z^5 e^z}{(e^z - 1)^2} dz. \qquad (4.15)$$

This expression has three fitting parameters: r_0 is the residual resistance due to impurity and boundary scattering of electrons; θ is a characteristic temperature depending on the material, e.g., θ for bulk crystalline gold is between 170 K and 200 K (see [6, Table 9.3]); and Δ is a scaling factor indicating the strength of electron–phonon coupling. Obtaining best-fit values for $r_{0,\text{fit}}$, θ_{fit}, and Δ_{fit}, is viable using standard software packages to minimize the residual (e.g., fminsearch in Matlab), and $m_{\text{B–G},0}(T)$ follows readily by analytically taking the temperature derivative of Eq. (4.15), as given in Eq. (4.16). However, for even a modestly complicated function such as Eq. (4.15) we are unaware of any general analytical approach to determining the *uncertainties* $u_{r_0}, u_\theta, u_\Delta$, and especially $u_{m_{\text{B–G}}}(T)$.

On the other hand, the MC scheme summarized in Fig. 4.4 is just as easy to implement for the nonlinear relation Eq. (4.15) as for the straight-line function $R_{e,\text{High-}T} = r_{0,\text{High-}T} + m_{\text{linear}}T$. We follow the recipe from Sec. 4.4, where now the control variable ($\mathbf{X}$) is a vector composed of the 22 temperatures (T_{set}) set in the cryostat; the response variable ($\mathbf{Y}$) is a vector composed of the 22 electrical resistances ($R_{e,\text{msrd}}$) measured at those temperatures; and the physics model is the Bloch–Grüneisen formula of Eq. (4.15) with its corresponding model parameter vector $\mathbf{a} = \{r_0, \theta, \Delta\}$. We also use our experimental judgment of the uncertainty distributions $\mathbf{u}_T = \pm 0.01$ K and $\mathbf{u}_R = \pm 0.01$ Ω, to feed into steps 4 and 6 of the recipe.

Following step 1 of the recipe in Fig. 4.4, we obtain $\mathbf{a}_0 = \{0.88\ \Omega, 158.9\ \text{K}, 1.29\ \Omega\}$ based on the complete calibration dataset $(T_{\text{set}}, R_{e,\text{mrsd}})$ from columns 1 and 2 of Table 4.2.

Then following steps 4–8 of the recipe with $N = 80$, we obtain a population $\{\mathbf{a}_1, \mathbf{a}_2, \ldots, \mathbf{a}_N\}$. For each $\mathbf{a}_i$ vector, we obtain an estimate for the slope

$$m_{\text{B–G},i}(T) = \frac{d}{dT}[R_{e,\text{B–G},i}(T)]$$

$$= \frac{4\Delta_i}{\theta_i}\left[5\left(\frac{T}{\theta_i}\right)^4\int_0^{\theta_i/T}\frac{z^5 e^z}{(e^z - 1)^2}dz - \left(\frac{\theta_i}{T}\right)^2\frac{e^{\theta_i/T}}{(e^{\theta_i/T} - 1)^2}\right].$$

$$(4.16)$$

At a specific temperature, e.g., 310 K, we evaluate Eq. (4.16) $N = 80$ times to obtain 80 different $m_{\text{B–G},i}(T = 310\,\text{K})$ based on the 80 $\mathbf{a}_i$ vectors. Following step 9 of the recipe, to evaluate the 95% CI we throw out the first and last points of the sorted 80 $m_{\text{B–G},i}(T = 310\,\text{K})$. The resulting lower bound $(m_{\text{B–G,LB}} = 8.21 \times 10^{-3}\ \Omega/\text{K})$ and upper bound $(m_{\text{B–G,UB}} = 8.30 \times 10^{-3}\ \Omega/\text{K})$ are listed in columns 6 and 7 in Table 4.2. Note that the slope, $m_{\text{B–G},0} = 8.26 \times 10^{-3}\ \Omega/\text{K}$ (column 5 in Table 4.2), is still calculated based on the initial fit, $\mathbf{a}_0$. Thus, we have found the 95% CI $[m_{\text{B–G,LB}}, m_{\text{B–G,UB}}]$, which may also be presented as $u_{m_{\text{B–G}}} = (m_{\text{B–G,UB}} - m_{\text{B–G,LB}})/2 = 4.67 \times 10^{-5}\ \Omega/\text{K}$ (column 8 in Table 4.2).

We end this section by comparing the linear fit and the Bloch–Grüneisen fit (Fig. 4.6). The two slopes differ by up to 8% over the 80–310 K temperature range, with the linear fit (blue) falling in between the band of the Bloch–Grüneisen fit (red). This indicates that we will both over- and under-predict the slope if we insist on a linear fit.

References

[1] W. Jang, Z. Chen, W. Bao, C. N. Lau and C. Dames, "Thickness-dependent thermal conductivity of encased graphene and ultrathin graphite," *Nano Lett.* **10**(10), pp. 3909–3913, 2010.

[2] S. Lee and D. G. Cahill, "Heat transport in thin dielectric films," *J. Appl. Phys.* **81**(6), pp. 2590–2595, 1997.

[3] A. D. McConnell and K. E. Goodson, "Thermal conduction in silicon micro- and nanostructures," *Annu. Rev. Heat Transf.* **14**, pp. 129–168, 2005.

[4] Z. Chen, W. Jang, W. Bao, C. N. Lau and C. Dames, "Thermal contact resistance between graphene and silicon dioxide," *Appl. Phys. Lett.* **95**(16), pp. 161910–161913, 2009.

[5] W. H. Press, S. A. Teukolsky, W. T. Vetterling and B. P. Flannery, *Numerical Recipes*, 3rd ed. New York: Cambridge University Press, 2007.

[6] J. M. Ziman, *Electrons and Phonons*. New York: Oxford University Press, 1960.

[7] S. J. Kline and F. A. McClintock, "Describing the uncertainties in single sample experiments," *Mech. Eng.* **75**(1), pp. 3–8, 1953.

[8] S. Pandya, J. D. Wilbur, B. Bhatia, A. Damodaran, C. Monachon, A. Dasgupta, W. P. King, C. Dames and L. W. Martin, "Direct measurement of pyroelectric and electrocaloric effects in thin films," *Phys. Rev. Appl.* **7**, p. 34025, 2017.

[9] A. J. Schmidt, X. Chen and G. Chen, "Pulse accumulation, radial heat conduction, and anisotropic thermal conductivity in pump-probe transient thermoreflectance," *Rev. Sci. Instrum.* **79**(11), p. 114902, 2008.

[10] W. Navidi, *Statistics for Engineers and Scientists*, 1st ed. Boston: McGraw Hill, 2006.

[11] D. C. Montgomery and G. C. Runger, *Applied Statistics and Probability for Engineers*, 3rd ed. John Wiley & Sons, 2003.

Appendices

These appendices cover a variety of secondary issues which also contribute to a successful experiment. We first introduce the lock-in amplifier, an important instrument which may be unfamiliar to some readers. Next we discuss the effect of natural convection on the 3ω method, followed by highlighting the advantages of a four-probe AC measurement, and offering an op-amp circuit to convert a voltage source to a current source. Then we determine the vacuum level needed in order to greatly suppress air conduction, and radiation shields to minimize radiation losses. At last, we discuss material properties for thermal design, and give a brief introduction to the lognormal distribution. A summary of the notation used in the book is also included.

A.   Lock-in amplifier [1]

The lock-in amplifier is a workhorse instrument used to detect small sinusoidal voltage signals at a known frequency. It functions essentially as a highly adjustable band-pass filter combined with an amplifier. The key feature of this filter is its extremely narrow bandwidth, with a quality factor ($f_s/\Delta f_{\text{bandwidth}}$, where f_s is the signal frequency and $\Delta f_{\text{bandwidth}}$ is the bandwidth) which can readily exceed 10^5.

In the following sections, we first outline the working principle of a lock-in amplifier and then highlight some of the key functions commonly used in thermal measurements at the nanoscale.

A.1. *Principle*

As shown in Fig. A.1, the input signal is composed of a signal of interest oscillating at an angular frequency ω_s, and background noise at all frequencies:

$$V_{\text{in}} = V_s \sin(\omega_s t + \phi_s) + \text{noise.} \qquad (A.1)$$

To detect this signal of interest, the lock-in uses a concept known as phase sensitive detection (PSD). This begins with the lock-in generating its own reference signal,

$$V_{\text{ref}} = V_L \sin(\omega_L t + \phi_L). \qquad (A.2)$$

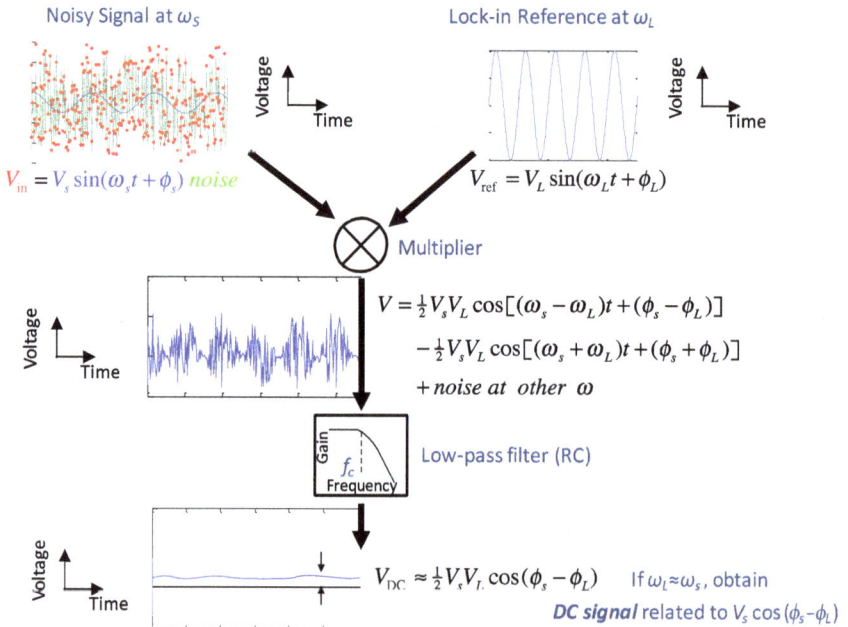

Fig. A.1 Principle of lock-in amplifier. Information involving the signal's amplitude and phase, $V_s \cos(\phi_s - \phi_L)$, is extracted by a phase sensitive detector (PSD), which is a combination of a multiplier and a low-pass filter. The depicted schematic gives the in-phase part of V_s. The out-of-phase part can be determined by sending V_{in} into a second PSD driven by another reference signal which has a $90°$ phase shift compared to ϕ_L (not shown). For details see main text.

The lock-in multiplies these two signals, giving

$$V = V_{\text{in}} \cdot V_{\text{ref}}$$
$$= \frac{1}{2} V_s V_L \cos[(\omega_s - \omega_L) \cdot t + (\phi_s - \phi_L)]$$
$$- \frac{1}{2} V_s V_L \cos[(\omega_s + \omega_L) \cdot t + (\phi_s + \phi_L)] + \text{noise at other } \omega.$$

$$(\text{A.3})$$

To measure the signal at ω_s, the lock-in reference must be set so that $\omega_L = \omega_s$, which is accomplished by phase locking the experiment to the reference, thus giving rise to the name "lock-in" amplifier. Then sending the multiplied signal through a low-pass filter of bandwidth $\Delta f_{\text{bandwidth}} \ll 2\omega_L/2\pi$, what remains is the DC component

$$V_{\text{DC}} = \frac{1}{2} V_s V_L \cos(\phi_s - \phi_L), \qquad (\text{A.4})$$

plus a residual contribution from the original noise contained in V_{in} in the range $\omega_L \pm 2\pi f_c$. Usually $\Delta f_{\text{bandwidth}}$ can be set small enough such that this residual noise is negligible (see Appendix A.3 below). Thus, since V_L is known, we can extract $V_s \cos(\phi_s - \phi_L)$. This represents the component of V_s which is in-phase with the lock-in reference.

Note that virtually all commercial lock-in amplifiers include two phase sensitive detectors (PSDs), which are phase shifted by 90° and thus record two DC components

$$V_{\text{PSD1}} = \frac{1}{2} V_s V_L \cos(\phi_s - \phi_L),$$
$$V_{\text{PSD2}} = \frac{1}{2} V_s V_L \sin(\phi_s - \phi_L).$$

$$(\text{A.5})$$

The in-phase and out-of-phase components of the original signal, often designated X and Y, respectively, are

$$X = \frac{V_{\text{PSD1}}}{\frac{1}{2} V_L} = V_s \cos(\phi_s - \phi_L),$$
$$Y = \frac{V_{\text{PSD2}}}{\frac{1}{2} V_L} = V_s \sin(\phi_s - \phi_L),$$

$$(\text{A.6})$$

which is what the lock-in reports. X and Y correspond to the real and imaginary parts of the solution of the 3ω methods (e.g., red and blue symbols in Fig. 2.3).

In 3ω measurements, the driving current is either directly taken from, or synchronized to, the reference signal output of the lock-in amplifier (see Fig. 2.2). Thus the phase difference, $\phi_s - \phi_L$, is not arbitrary but instead is determined by the thermal transfer function (Z_ω): if Z_ω is a pure real function, $\phi_s - \phi_L = 0$, and thus there is only X but no Y component; if Z_ω is a complex function (e.g., in a typical 3ω configuration), $\phi_s - \phi_L \neq 0$, and thus there are both X and Y components.

A.2.   *Full scale sensitivity and least significant bit (LSB)*

While early lock-in amplifiers implemented the process of Fig. A.1 entirely in hardware, in modern instruments it is now more common to immediately digitize the input signal $V_{in}(t)$ and perform all subsequent steps in software. In this case, the least significant bit (LSB) is an important consideration determining the measurement detection limits. For example, an SR850 digitizes the input signal using an analog-to-digital converter with $N = 14$ bits, and the LSB is related to the full scale sensitivity (FS) by

$$\text{LSB} = \frac{\text{FS}}{2^N}. \tag{A.7}$$

Figure A.2 shows an example in which the LSB is not adequate to detect the difference between the adjacent data points. An ideal measurement with correct FS settings should generate a smooth linear relation between $V_{3\omega}$ and the logarithm of the driving frequency, as shown in the red symbols in Fig. 2.3. However, the data depicted in Fig. A.2 show clear stepwise discretization artifacts spaced by $\sim 0.5\ \mu V$, comparable to the LSB of $0.6\ \mu V$. The appearance of such prominent discretization artifacts as in Fig. A.2 strongly suggests that the FS should be reduced if possible without saturating the input of the lock-in.

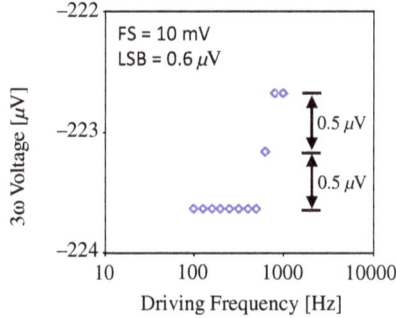

Fig. A.2 An example of discretization artifacts due to inadequate LSB of lock-in amplifier for a 3ω measurement. The true voltage vs. frequency response should be a smooth continuous function. However, stepwise voltage artifacts are clearly observed, comparable to the 0.6 μV LSB which comes from an FS of 10 mV and 14 bits of discretization.

Knowing the number of bits, N, we can estimate the resolution of a resistance thermometer detected by a lock-in amplifier. Assuming no background subtraction and that the FS value is set optimally, the detectable fractional change of electrical resistance can be related to N by

$$\frac{\delta R_e}{R_e} \sim \frac{1}{2^N}, \tag{A.8}$$

and thus detectable change of temperature is

$$\delta T = \frac{\delta R_e}{R_e} \frac{1}{\alpha}$$

$$\simeq \frac{1}{2^N} \frac{1}{\alpha}, \tag{A.9}$$

where α is the temperature coefficient of the resistor.

Example A.1. Estimate the resolution of a temperature sensor made of gold and monitored by an SR850 lock-in amplifier in the vicinity of room temperature.

Solution: We use the literature value, $\alpha_{Au} \sim 3.4 \times 10^{-3}/$K (see [2, p. 602]); for an SR850, as stated above we have $N = 14$. Thus, using Eq. (A.9), we estimate the resolution to be $\delta T \sim 0.02$ K.

Note that if the sensor is microfabricated, such as the heater pattern in a typical 3ω measurement, α could be substantially lower than the literature value (see Appendix G.4).

If finer temperature resolution is desired below one LSB, one effective strategy is to use a Wheatstone bridge or related background subtraction prior to digitization by the lock-in, which when carefully implemented can reduce the detection limit to below $\sim$100 μK [3, 4]. Surprisingly, in principle one can also benefit by deliberately making the input signal *noisier* so that it stochastically samples multiple bit levels, followed by subsequent averaging. This concept, known as "stochastic resonance", has been applied to optical thermoreflectance measurements [5], though not to our knowledge for resistance thermometry.

A.3. *Time constant*

The most important parameter characterizing the low-pass filter of Fig. A.1 is its time constant, which can be defined as

$$\tau_e = \frac{1}{2\pi f_c}, \tag{A.10}$$

where f_c is the -3 dB frequency of the low-pass filter's transfer function.

The two main issues involving the time constant are as follows:

(i) What τ_e should we choose?
(ii) After making a change to the experimental conditions, how long is the stabilization time, in multiples of τ_e, required for the signal to reach its new steady state?

For the first question, a larger τ_e gives a smaller f_c, which as noted below Eq. (A.4) is linearly proportional to the passband for noise sources with frequencies close to ω_L to pass through to the final V_{DC} and cause errors. Thus, the larger the τ_e, the more accurate and stable the final measurement. It is also generally a good idea to ensure $\omega_L \tau_e > 1$.

However, there is also a downside to using large τ_e because it slows down the time to respond to a step change, as seen in the following answer to question (ii). Let us analyze the low-pass filter, an elementary RC circuit. Recall the capacitor discharging process from an initial V_0:

$$V(t) = V_0 \exp\left(-\frac{t}{\tau_e}\right). \tag{A.11}$$

Let us estimate the time required for the most extreme step: from $V_0 \approx V_{FS}$ down to the practical zero level $V(t_{\text{stabilize}}) \approx V_{LSB}$. We have

$$V_{LSB} = V_{FS} \exp\left(-\frac{t_{\text{stabilize}}}{\tau_e}\right). \tag{A.12}$$

Thus, the time required to stabilize is

$$t_{\text{stabilize}} = \tau_e \ln \frac{V_{FS}}{V_{LSB}}$$
$$= \tau_e \ln 2^N$$
$$\approx 10\tau_e, \tag{A.13}$$

where the last step used $N = 14$ for an SR850.

Likewise, in a charging process, we replace Eq. (A.11) with

$$V(t) = V_0 \left[1 - \exp\left(-\frac{t}{\tau_e}\right)\right], \tag{A.14}$$

and for a step from 0 up to $V(t_{\text{stabilize}}) \approx V_{FS} - V_{LSB}$, we replace Eq. (A.12) with

$$V_{FS} - V_{LSB} = V_{FS} \left[1 - \exp\left(-\frac{t_{\text{stabilize}}}{\tau_e}\right)\right]. \tag{A.15}$$

Again, we obtain the same expression as Eq. (A.13).

Equation (A.13) shows how the time to stabilize a voltage change to within one LSB is directly proportional to τ_e, with a multiplier of ~ 10 for the specific example of an SR850. Based on our empirical experience, to be conservative we typically use a multiplier of

$t_{\text{stabilize}}/\tau_e \approx 15 - 20$ in a practical measurement. This rapidly becomes a major inconvenience for $\tau_e > 10$ s.

Summarizing the competing effects on τ_e, as a rule of thumb $\tau_e \sim 0.3$ s–1 s is a good starting point in a practical measurement. However, τ_e must be increased in certain circumstances. For example, in a 3ω measurement of a material with low thermal diffusivity, if a large penetration depth is desired it may be necessary to drive the heater with low frequencies, e.g., ~ 0.1 Hz, in which case we must increase $\tau_e > 1$ s.

A.4. *Dynamic reserve*

A lock-in amplifier's dynamic reserve (DR) relates the maximum tolerable noise signal ($V_{\text{max-tolerable-noise}}$) to the full-scale signal (V_{FS}) as

$$DR = 20 \log_{10} \left(\frac{V_{\text{max-tolerable-noise}}}{V_{\text{FS}}} \right), \qquad (A.16)$$

which has units of dB. For example, if we set a full-scale sensitivity of $V_{\text{FS}} = 10$ mV, and DR $= 30$ dB, we would be able to tolerate a noise background as large as

$$V_{\text{max-tolerable-noise}} = V_{\text{FS}} \cdot 10^{\text{DR}/20} \simeq 316\,\text{mV}.$$

Here "noise" must be understood to include any background signal, whether broadband or a smooth sine wave. The latter is especially relevant for 3ω measurements, because the small 3ω signals are accompanied by 1ω voltages which are routinely $100 \times$ to $1000 \times$ larger. Although this large 1ω background is most commonly dealt with using a separate hardware subtraction circuit (e.g., see [6, Fig. 4]), in special cases it has been shown possible to forego the background subtraction circuit by taking advantage of a carefully-chosen dynamic reserve setting in a lock-in amplifier [7].

One caution is that it can also be problematic if the DR is set too high. One issue is that high DR also increases the detection limit V_{LSB}, degrading the accuracy of small signal measurements. At ultra high reserves, even the lock-in's own output noise may become detectable (see [1, pp. 3–18]).

Taking into account the above considerations, a suggested routine in a practical measurement, e.g., measuring the 3ω voltage without any background subtraction, is as follows. Starting with a fairly large DR, we first set the FS to be just larger than the largest expected amplitude of $V_{3\omega}$. Then we gradually reduce the DR to find a minimum DR which can still tolerate the background noise including 1ω background without overloading the lock-in's input. Even for a constant input signal measured without overloading, changing the FS and DR settings can also slightly change the reported voltages. Therefore, we find best practice is to hold the FS and DR settings constant for all $V_{3\omega}$ measurements over the entire range of frequencies, 1ω currents, and temperatures studied for a given sample. This requires some forethought about the extreme values of $V_{1\omega}$ and $V_{3\omega}$ expected, so the FS and DR can be set to accommodate all scenarios.

A.5. *Johnson noise*

Johnson noise is a fundamental result of thermal fluctuations. For large resistors, we can detect this Johnson noise using a lock-in amplifier. The root-mean-square (RMS) value of a voltage generated by Johnson noise can be expressed as

$$V_{\text{Johnson-noise}} = \rho_{\text{Johnson-noise}} \cdot \sqrt{f_{\text{ENBW}}}, \tag{A.17}$$

where f_{ENBW} is the equivalent noise bandwidth (ENBW) of the lock-in amplifier, which can be linked to the time constant (τ_e) as

$$f_{\text{ENBW}} = \frac{1}{4\tau_e} \tag{A.18}$$

for a single stage low-pass RC filter with 6 dB/octave roll off. Note here f_{ENBW} is not exactly the same as $f_c = 1/2\pi\tau_e$ in Eq. (A.10), differing by a factor of 1.57. The noise density $\rho_{\text{Johnson-noise}}$ is defined as

$$\rho_{\text{Johnson-noise}} = \sqrt{4k_B T R_e}, \tag{A.19}$$

where k_{B} is the Boltzmann constant, T is the absolute temperature, and R_e is the electrical resistance of the resistor. For example, at room temperature we have $\rho_{\text{Johnson-noise}}$ = 0.91, 4.1, and 130 nV/$\sqrt{\text{Hz}}$ for a 50 Ω, 1 kΩ, and 1 MΩ resistor, respectively.

A representative lock-in amplifier, the SR850, has a specified input-referred noise floor of 5 nV/$\sqrt{\text{Hz}}$ (see [1, pp. 3–19]). Thus, with $\tau_e = 0.2$ s so that $f_{\text{ENBW}} = 1.25$ Hz, any external noise (or indeed signal) higher than 5.6 nV should be detectable; otherwise it will be overwhelmed by the instrument's own noise. This is consistent with the measurements in Fig. A.3: while $V_{\text{Johnson-noise}} = 1.0$ and 4.5 nV for the 50 Ω (green solid line) and 1 kΩ (pink symbol) resistors are entangled with the input noise floor (black dashed line), $V_{\text{Johnson-noise}} = 142$ nV for the 1 MΩ resistor (blue solid line) is clearly distinguished.

This Johnson noise voltage has also been used as a thermometer to detect the temperature of a resistor of known resistance, although this requires more sophisticated hardware [8, 9].

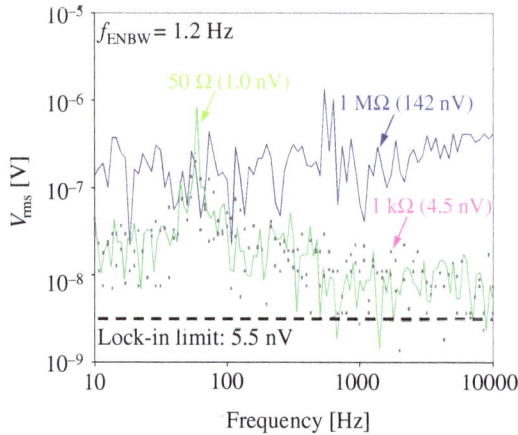

Fig. A.3 Voltage noise from three different resistors, connected directly across the measurement terminals of a lock-in amplifier without any driving current source. The voltages in parentheses are the corresponding Johnson noise values for each of the three resistors. While the root-mean-square voltage (V_{rms}) noise for the 50 Ω (green solid line) and 1 kΩ (pink symbol) resistors are down around the lock-in's own noise floor (black dashed line), especially at high frequencies, the larger Johnson noise from the 1 MΩ resistor (blue solid line) is clearly above the floor.

B.   Effect of natural convection on the 3ω method

We may wonder how natural convection would affect the results of the 3ω method in a standard laboratory environment, especially when we are not equipped with a high-vacuum cryostat, or when we want to conduct a quick and dirty measurement outside the cryostat at room temperature.

We consider a typical scenario, a line heater on top of a solid substrate (Fig. B.1(a)). The corresponding thermal circuit is shown in Fig. B.1(b): heat leaks parasitically upwards through natural convection, in addition to the preferred pathway of conduction down through the solid.

In order to quantify the effect of natural convection, we can introduce a Biot number,

$$Bi = \frac{R_{\text{cond'n}}}{R_{\text{conv'n}}}, \tag{B.1}$$

where $R_{\text{cond'n}}$ and $R_{\text{conv'n}}$ are the thermal resistance of the bottom solid and the top air, respectively. Only if this $Bi \ll 1$ can the losses through natural convection be safely neglected.

We now model $R_{\text{conv'n}}$ and $R_{\text{cond'n}}$, focusing on the in-phase response. First, rearranging Eq. (2.13), we obtain the conductive

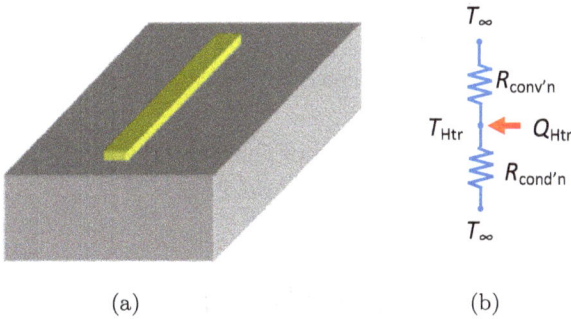

(a)   (b)

Fig. B.1   A typical heater-on-substrate configuration (a), and a corresponding simplified thermal circuit (b), to analyze the effect of natural convection on a measurement of thermal conductivity.

thermal resistance through the bottom solid as

$$R_{\text{cond'n}} = \frac{1}{\pi l k_{\text{sub}}} \left[\ln \left(\frac{L_p}{b} \right) + \eta \right],$$

(B.2)

where $\eta \approx 0.923$ [10].

As discussed with Eqs. (2.13) and (2.14), Eq. (B.2) represents a modification of a DC cylindrical heating problem (see [11, p. 69]), to a hemi-cylinder, with the inner radius as the half-width of the line heater, b, and the outer radius as the penetration depth, $L_p = \sqrt{D_{\text{sub}}/\omega_H}$, in which D_s is the thermal diffusivity, and $\omega_H = 2 \cdot \omega$ is the frequency of the periodic heating.

Next, we estimate the convective thermal resistance through the upper air as

$$R_{\text{conv'n}} = \frac{1}{h(l \cdot 2L_p)},$$

(B.3)

in which we approximate the heated surface area of the top of the substrate as a rectangle of length l and width $2L_p$.

Combining the three equations above, we obtain

$$Bi = \frac{h \cdot 2L_p}{\pi k_{\text{sub}}} \left[\ln \left(\frac{L_p}{b} \right) + \eta \right].$$

(B.4)

Recalling the typical analysis of the convective heat transfer, we have the Nusselt number defined as

$$Nu_D \equiv \frac{hD}{k_{\text{air}}},$$

(B.5)

where D is the effective breadth of the heated zone, which physically should be somewhere between $2b$ and $2L_p$.

Thus, we have

$$Bi = \frac{Nu_D}{\pi} \frac{k_{\text{air}}}{k_{\text{sub}}} \frac{2L_p}{D} \left[\ln \left(\frac{L_p}{b} \right) + \eta \right].$$

(B.6)

Now the key is to estimate the Nusselt number, Nu_D. As an order of magnitude estimate, we consider natural convection from horizontal isothermal cylinders, with a Nusselt number (see [11, p. 418])

$$Nu_D = 0.36 + \frac{0.518 Ra_D^{1/4}}{[1 + (0.559/\text{Pr})^{9/16}]^{4/9}}, \tag{B.7}$$

where $\text{Pr} \equiv \nu/D_{\text{air}}$ is the Prandtl number, and the Rayleigh number is defined as

$$Ra_D \equiv \frac{gD^3}{\nu D_{\text{air}}} \frac{\Delta T}{T_\infty}, \tag{B.8}$$

where g is the gravitational constant, and ν and D_{air}, respectively, are the kinematic viscosity and thermal diffusivity of the surrounding air.

Combining Eqs. (B.6)–(B.8), we have $Bi \propto D^{-n}$, where n is between $1/4$ to 1. For a conservative estimate, here we consider our smallest estimate for D, approximating $D \approx 2b$ rather than $2L_p$ (Sec. 2.2.1).

For a typical microfabricated line heater with $D = 2b \sim 10\,\mu\text{m}$, and a preferred $\Delta T/T_\infty \sim 1\%$, for air at room temperature Eq. (B.7) is dominated by its first term. Substituting this into Eq. (B.6), we have

$$Bi = 0.11 \frac{k_{\text{air}}}{k_{\text{sub}}} \frac{L_p}{b} \left[\ln\left(\frac{L_p}{b}\right) + 0.923 \right]. \tag{B.9}$$

For a conservative estimate (erring on the side of large Bi, since small Bi is desired to neglect convection), we consider a poor thermal conductor, such as SiO_2, and a low frequency (1 Hz), which gives $L_p/b \sim 50$. The result is $Bi < 1 \times 10^{-2}$. Thus, since $Bi \ll 1$, we can safely neglect the parasitic heat losses through natural convection at room temperature.

We close by commenting on two shortcomings of the analysis above. First, Eq. (B.7) is for a freestanding cylinder. In the scenario of Fig. B.1, the convection coefficient, h, should be significantly reduced, because the surrounding substrate impedes the fluid flow as compared to a free cylinder. This overestimation of h means that the above analysis was conservative (tending to overestimate Bi). Second, note also that Eq. (B.7) was stated to be valid only for $Ra_D \geq 1 \times 10^{-6}$. However, here we have $Ra_D \approx 0.3 \times 10^{-6}$ at room temperature for the conservative estimate $D = 2b$.

Considering the very large factor of safety in Bi just estimated, it is still reasonable to use Eq. (B.7) for an order-of-magnitude estimate.

C.   Advantages of a 4-point probe AC measurement

The accuracy of a resistance thermometer depends on the accuracy of the resistance measurement at various temperatures. Figure C.1 shows a schematic (left) and the corresponding circuit (right) of three different schemes to measure the electrical resistance of a microfabricated resistive thermometer.

The two-probe DC method depicted in Fig. C.1(a) is not uncommon in general-purpose resistance measurements in the lab, for example, using a Digital Multi-Meter (DMM) in two-wire mode. However, for precision resistance thermometry applications such a two-probe method may introduce large errors, especially if the experiment involves long leads or low temperatures.

There are two major error sources in Fig. C.1(a). The first one is the Seebeck effect. It is very common that the lead wires and the micro-thermometer are made of two different materials, for example copper and gold, thereby forming a thermocouple at every junction. If temperature differs between these junctions, a net thermal EMF is generated (e.g., V_{s1} not canceling V_{s2} in Fig. C.1(a)). The DC voltmeter cannot distinguish these undesired thermal EMFs from the desired ohmic voltage across the resistive thermometer $R_{e,\text{sampl}}$. Because the Seebeck voltage scales with temperature difference, these errors tend to be largest in experiments involving large temperature differences, such as between room temperature and a cryogenic sample stage, or room temperature and a high temperature stage. Compared to metals, the Seebeck coefficient also can be much larger in semiconductors and some contact pastes, which can also increase these errors if they are part of the circuit path.

The second source of error in Fig. C.1(a) comes from the electrical resistance of the long leads and the electrical contact resistances at the lead-thermometer junctions. In the circuit in Fig. C.1(a) (right), we combined these two components and labeled them as $R_{e,\text{LC1}}$ and

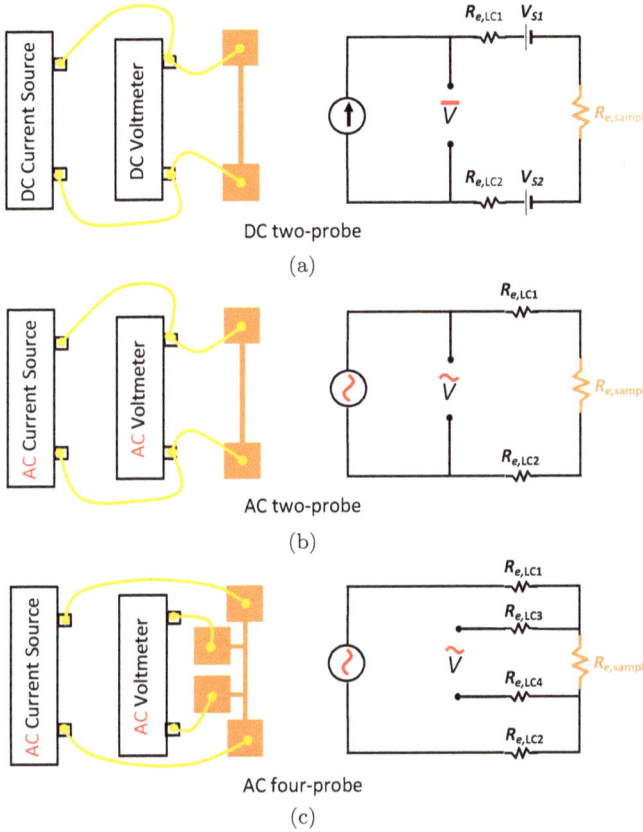

Fig. C.1 Advantages of a four-probe AC measurement. (a) A conventional two-probe measurement driven by a direct current suffers from two error sources: thermoelectric EMFs at the junctions between dissimilar metals (represented by V_{s1} and V_{s2}), and ohmic voltage drops across long leads and at the contacts (represented by $R_{e,\text{LC1}}$ and $R_{e,\text{LC2}}$). (b) An AC measurement is immune from the thermoelectric EMFs. (c) A four-probe AC measurement is immune from both error sources. In practice, the AC voltages should be measured with a lock-in amplifier, which has much better sensitivity and frequency selectivity than a general purpose digital multimeter.

$R_{e,\text{LC2}}$. Electrical current flowing through these resistances generates undesired voltages, which again the voltmeter cannot separate out from the desired voltage drop across R_{sampl}. Clearly $R_{e,\text{LC}}$ becomes larger when the leads are thin, long, and made from metals of high electrical resistivity. One example is again cryogenic measurements, where long metallic leads with low thermal conductivity are sometimes selected in order to reduce parasitic heat losses by conduction. However, for such metal wires, low thermal conductivity comes with high electrical resistivity due to the Wiedemann–Franz law (Eq. (3.26)).

To eliminate the errors caused by thermal EMF, we switch from the DC measurement to an AC measurement (Fig. C.1(b)). The key here is the fact that the ohmic voltages follow the frequency of the driving current, while the temperature differences which cause the thermal EMFs are generally low-frequency drifts close to DC. Thus, by focusing only on the AC response, now the AC voltmeter can exclude the thermal EMF effects.

To eliminate the errors caused by the ohmic voltage drops along the long leads and contacts, we switch from two-wire measurement to four-wire measurement (Fig. C.1(c)). This separates the voltage probes from the current probes, and thus the AC voltmeter does not see these undesired voltage components. Although the voltage probe leads also have finite resistance ($R_{e,LC3}$ and $R_{e,LC4}$ in Fig. C.1(c)), there is virtually no current flowing through them, so those additional voltage drops are negligible.

To conclude, applying AC measurements in a four-probe configuration like Fig. C.1(c) is the standard for electro-thermal measurement methods as discussed in this book, because this eliminates errors from both thermal EMFs and the lead and contact resistances. It is strongly recommended that the "AC Voltmeter" function in Fig. C.1 be implemented with a lock-in amplifier rather than a general purpose DMM in AC volts mode, because the former can be locked specifically to the frequency of the AC current source with a very tight bandwidth and has much better sensitivity and frequency selectivity. These features allow a lock-in to exclude additional noise sources such as 50 Hz/60 Hz from the power lines and various artifacts related to joule heating, which occurs at twice the frequency as the driving current.

Example C.1. Constantan is sometimes used for lead wires inside a cryostat, due to its low thermal conductivity and low temperature coefficient of resistivity. However, constantan is also a common material for thermocouples, due to its fairly high (negative) Seebeck coefficient.

(a) Look up the electrical resistivity of constantan at room temperature, and estimate the corresponding electrical resistance for a 1-m long lead wire with a diameter of 5 mm. How reasonable would it be to neglect this R_L as compared to that of a typical microfabricated thermometer ($R_{sampl} \sim 10 \ \Omega$)?

(b) Similarly, find the Seebeck coefficients of constantan and copper, and estimate the net thermal EMF arising from a constantan lead which connects a copper thermometer at 600 K to additional copper wiring at room temperature. Can we neglect this voltage when using sensing currents in the range 0.1–1 mA?

Solution: (a) The resistivity of constantan at room temperature is $\rho = 4.9 \times 10^{-7} \ \Omega \cdot m$ (see [12, p. 29]), and so the wire's electrical resistance is $R_L = \rho L / A = 39 \ \Omega$. Note that this is even larger than the typical R_{sampl}, and thus certainly cannot be neglected.

(b) The Seebeck coefficients of constantan and copper are $-50 \ \mu V/K$ and $3.5 \ \mu V/K$, respectively, averaged from room temperature to 600 K (see [12, p. 31]). The total thermal EMF of the junction is $V_{s1} - V_{s2} = (S_{constantan} - S_{copper}) \times (T_{junction1} - T_{junction2}) = 16 \ mV$. Note that this is even larger than the voltage generated by a typical sensing current of 0.1–1 mA through the 10 Ω sample and thus certainly cannot be neglected.

From this example, an AC, four-probe measurement is mandatory for accurate resistance thermometry.

D. Voltage to current conversion

The 3ω method assumes the sample is driven by an ideal sinusoidal current source, and the resulting 1ω and 3ω voltages are measured using a lock-in amplifier (see Fig. 2.2). Although commercial AC current sources have recently become available (e.g., Keithley 6221), for simplicity and lower cost it is appealing to drive the 3ω circuit

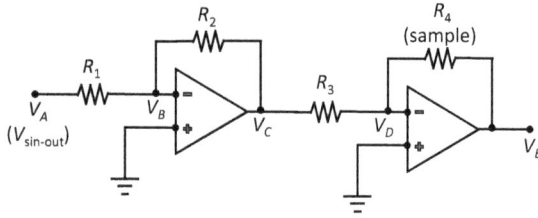

Fig. D.1 A voltage-to-current op-amp circuit. The input is a voltage ($V_{\text{sin-out}}$) provided by a lock-in amplifier, which is converted to a current through the sample (R_4), according to $I_4 = -\frac{R_2}{R_1 R_3} V_{\text{sin-out}}$. The key feature is that this sample current is independent of the sample resistance.

using the lock-in amplifier's own sine wave output. However, that output is a voltage source, typically with a nonnegligible 50 Ω output impedance. Although in principle the standard 3ω analysis can be corrected for a driving voltage source [7, 14], it may be more straightforward to instead convert the voltage source to a current source.

Figure D.1 presents a simple op-amp circuit [13] which converts the voltage reference from the lock-in amplifier ($V_{\text{sin-out}}$) to a current (I_4) flowing through the sample ($R_{e,4}$). The key feature is that this current

$$I_4 = -\frac{R_{e,2}}{R_{e,1} R_{e,3}} V_{\text{sin-out}} \tag{D.1}$$

does not depend on R_4.

Example D.1. Recall the two assumptions for an ideal op-amp:

(I) There is no current flowing in/out of the "+" and "−" terminals.
(II) A feedback forces $V_+ = V_-$.

Apply this ideal op-amp model to the schematic of Fig. D.1 to derive Eq. (D.1).

Solution: Applying the first rule to the left op-amp, we have

$$\frac{V_A - V_B}{R_{e,1}} = \frac{V_B - V_C}{R_{e,2}}.$$

Applying the second rule to the left op-amp, we have

$$V_B = 0.$$

Combining these two equations, we obtain

$$V_C = -\frac{R_{e,2}}{R_{e,1}}V_A.$$

Likewise, applying the two rules to the right op-amp, we have

$$V_E = -\frac{R_{e,4}}{R_{e,3}}V_C.$$

Now eliminating V_C from these last two equations, we obtain

$$V_E = \frac{R_{e,2}}{R_{e,1}}\frac{R_{e,4}}{R_{e,3}}V_A,$$

and thus

$$I_4 = \frac{0 - V_E}{R_{e,4}}$$

$$= -\frac{R_{e,2}}{R_{e,1}R_{e,3}}V_A$$

$$= -\frac{R_{e,2}}{R_{e,1}R_{e,3}}V_{\text{sin-out}}.$$

Note that V_E has dropped out of the analysis, though in practice it is connected through some low-resistance pathway to ground, e.g., to the negative terminal of the lock-in's sin-out port.

As a concrete example, we sometimes use Texas Instruments OPA551 for the op-amps, with $R_{e,1} = 1$ $k\Omega$, $R_{e,2} = 10$ $k\Omega$, and $R_{e,3} = 100$ Ω. In this case, if we set $V_{\text{sin-out}} = 1$ V, the corresponding current flowing through the sample ($R_{e,4}$) is 100 mA, with a 180° phase shift.

E. Cryostat and vacuum level

To minimize parasitic heat losses from convection, and to ensure a stable cryostat environment with minimal cryogen consumption, thermal measurements at the nanoscale are commonly conducted in vacuum. Correspondingly, one frequently asked question is whether a cheap roughing pump is adequate for this purpose, or if an expensive high-vacuum pump is needed.

Because radiation and air conduction are parallel pathways, let us analyze the requirements for the parasitic air conduction losses to be at least an order of magnitude smaller than those through radiation, since there is little benefit in further reducing the air conduction beyond that point.

We first estimate the relative importance of these two heat transfer modes. Referring to Fig. E.1, for air conduction we have a heat transfer coefficient

$$h_{\mathrm{cond'n}} = \frac{k_{\mathrm{air}}}{L_c}, \tag{E.1}$$

where k_{air} is the thermal conductivity of air and L_c is the characteristic length of the chamber, e.g., the gap between the sample stage and the chamber walls.

Recalling Eq. (3.6), for radiation driven by a small or moderate ΔT, where $\Delta T = T_{\mathrm{sampl}} - T_{\mathrm{chamber}}$ is the temperature difference between the sample and the chamber walls, we have (see [11, p. 75])

$$h_{\mathrm{rad'n}} = 4\varepsilon\sigma T_{\mathrm{avg}}^3, \tag{E.2}$$

where ε is the emissivity of the sample, $\sigma = 5.67 \times 10^{-8}\ \mathrm{W\,m^{-2}\,K^{-4}}$ is the Stefan–Boltzmann constant, and $T_{\mathrm{avg}} = (T_{\mathrm{sampl}} + T_{\mathrm{chamber}})/2$. This linearized expression is a reasonable approximation for ΔT up to $\sim T_{\mathrm{avg}}/2$. It also approximates the chamber walls as black and/or having a surface area much larger than the sample itself, and $h_{\mathrm{rad'n}}$

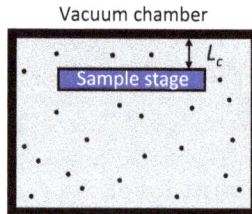

Fig. E.1 Schematic of parasitic heat losses between a sample stage at temperature T_{sampl} and the walls of the vacuum chamber at T_{chamber}. Here radiation and rarefied gas conduction are parallel heat transfer pathways.

could be greatly reduced further by surrounding the sample by one or more radiation shields (Appendix F).

As a concrete example, consider the schematic in Fig. E.1. Near room temperature, we have $h_{\mathrm{rad'n}} \approx 0.6$ W/m$^2 \cdot$K for a shiny metal sample stage, and $h_{\mathrm{cond'n}} \approx 3$ W/m$^2 \cdot$K for $L_c \sim 1$ cm. Based on these estimates, the conduction losses are larger than the radiation losses, so it would be worthwhile to evacuate the system to suppress the air conduction to $h_{\mathrm{cond'n}} \ll h_{\mathrm{rad'n}}$.

To explore how vacuum can suppress air conduction, we start from the well-known kinetic theory expression for the thermal conductivity

$$k = \frac{1}{3} C v \Lambda, \tag{E.3}$$

where C, v, and Λ are the volumetric heat capacity, velocity, and mean free path (MFP) of air molecules, respectively [15, 16]. The dominant phenomena can be understood by approximating air as a monatomic ideal gas, in which case these parameters scale with respect to air pressure as follows:

$$C \sim p^1, \tag{E.4a}$$

$$v \sim p^0, \tag{E.4b}$$

$$\Lambda \sim p^{-1}. \tag{E.4c}$$

Together, Eqs. (E.3) and (E.4) imply that the thermal conductivity does not depend on pressure (Fig. E.2(a)), indicating that there is no benefit whatsoever to applying a vacuum pump! However, this conclusion is true only for a system with infinitely large L_c. For a real system, Λ cannot exceed the finite L_c (Fig. E.2(b)); that is, at sufficiently high vacuum Λ is truncated to $\sim L_c$. This breaks the pressure scaling given in Eq. (E.4(c)), and instead $k \propto p$ at sufficiently high vacuum. This is the reason why pumping to high vacuum can indeed reduce $h_{\mathrm{cond'n}}$.

Finally, we derive a simple estimate for the vacuum level needed to reduce the gas thermal conductivity to a desired level. Consider pumping down a vacuum chamber from atmospheric conditions, corresponding to initial thermal conductivity, pressure, and MFP of k_0, p_0, and Λ_0, respectively. At first, there is no reduction in thermal

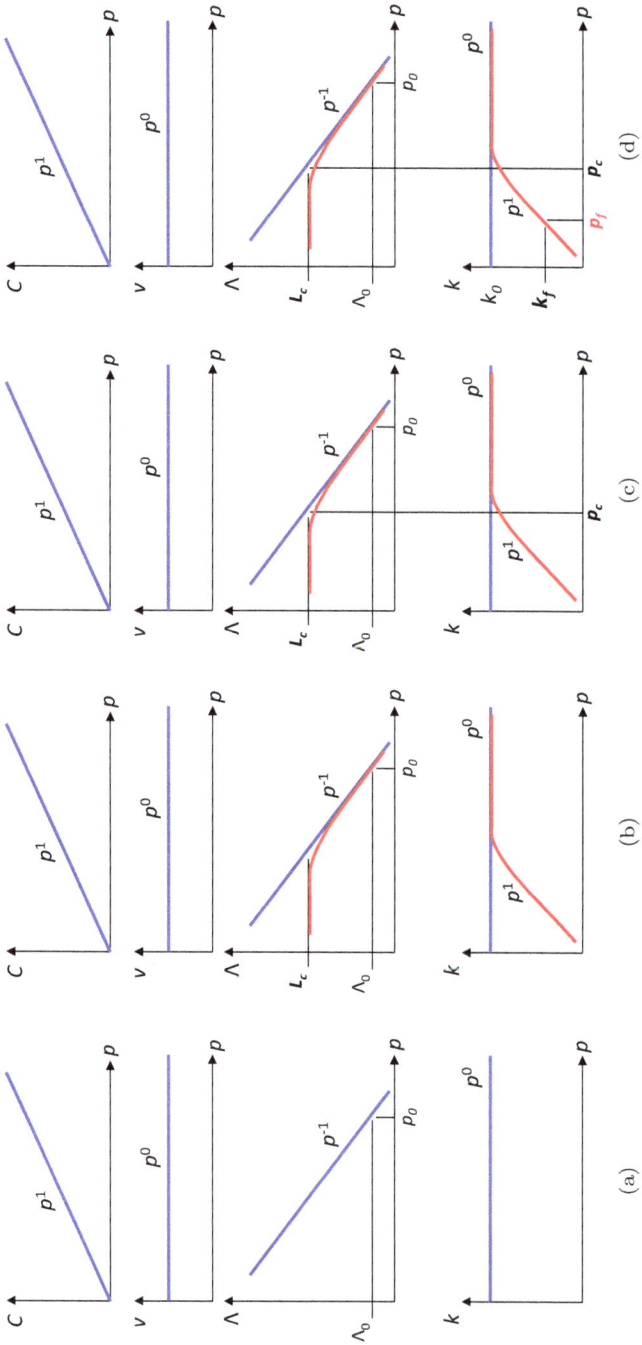

Fig. E.2 Thermal conductivity, k, of an ideal gas as a function of pressure, p, understood through kinetic theory. From top to bottom, the four rows depict the pressure dependence of C, v, Λ, and k, respectively, all on log–log axes. (a) In an ideal system of infinite size, k is independent of p. (b) In a realistic system with finite size (red lines), k does depend on p due to truncation of Λ by the characteristic length, L_c, such as the gap shown in Fig. E.1. Panels (c) and (d) are a graphical interpretation of finding a final pressure p_f to achieve a required low thermal conductivity k_f. (c) Focusing on the $\Lambda(p)$ plot, move from atmospheric pressure, p_0, to the characteristic transition pressure, p_c, at which $\Lambda \sim L_c$. (d) Then in the $k(p)$ plot, move from p_c to p_f using $k \propto p$. Combining (c) and (d) gives a guideline to the final pressure: $p_f = (\frac{\Lambda_0}{L_c})(\frac{k_f}{k_0})p_0$, where the first fraction on the right-hand side represents the initial pumping effort to reach the transition point, and the second fraction represents the additional pumping effort to reduce the thermal conductivity of the air.

conductivity, until some characteristic pressure, p_c, at which Λ has increased to $\sim L_c$. From Eq. (E.4(c)) and Fig. E.2(c), we can easily estimate this characteristic transition point as

$$p_c = \frac{\Lambda_0}{L_c} p_0. \tag{E.5}$$

Then during further pumping to a final pressure $p_f \ll p_c$, the system is in the rarefied gas regime of $k \sim p^1$ (Fig. E.2(d)), so that we have

$$p_f = \frac{k_f}{k_c} p_c, \tag{E.6}$$

where $k_c(\approx k_0)$ is the thermal conductivity at the transition point, and k_f is the target thermal conductivity.

Combining Eqs. (E.5) and (E.6), we have

$$p_f = \left(\frac{\Lambda_0}{L_c}\right)\left(\frac{k_f}{k_0}\right) p_0 \quad (p \ll p_c). \tag{E.7}$$

This form has been written to emphasize the two main phenomena: the first grouping on the right-hand side represents the initial pumping effort to reach the characteristic transition p_c, and the second grouping represents the additional effort to reduce the thermal conductivity of the air.

Example E.1. Consider air at 1 atm and 300 K, and take its MFP to be $\Lambda_0 \approx 150$ nm (see [16, p. 28]). (a) For a characteristic length of

$L_c = 1$ cm, estimate the characteristic pressure to transition between $k = $ const. and $k \propto p$ regimes. (b) What vacuum level is needed to suppress the air conduction by a factor of 100?

Solution: (a) From Eq. (E.5), $p_c = (\frac{150\,\mathrm{nm}}{1\,\mathrm{cm}}) \times 1\,\mathrm{atm} = 1.5 \times 10^{-5}$ atm $= 1.1 \times 10^{-2}$ Torr. This may be reached with some roughing pumps.

(b) From Eq. (E.6) with $k_f/k_0 = 0.01$, we need to reduce the pressure by another factor of 100 below p_c, and thus

$$P_f \sim 1.1 \times 10^{-4} \text{ Torr.}$$

This is too low for a roughing pump, so we will need a high vacuum pump.

F. Radiation shields

In Sec. 3.2.1, we concluded that radiation losses from the top surfaces of the microfabricated electrodes and the sample are usually negligible at and below room temperature. However, this might not be the case for high temperature measurements since $h_{\mathrm{rad'n}}$ could be enhanced from 6.1 W/m$^2 \cdot$K to 227 W/m$^2 \cdot$K (see Eq. (3.6)) if increasing the average temperature from 300 K to 1000 K, which may no longer be negligible as compared to $h_{\mathrm{cond'n}}$ (Eq. (3.14)).

If the radiation loss is deemed unacceptably high, a common strategy to reduce it is to use one or more radiation shields. For N concentric shields the radiative heat transfer can be expressed as [17]

$$Q = \frac{\sigma(T^4 - T_\infty^4)}{\left(\frac{1-\varepsilon}{\varepsilon A} + \frac{1}{A} + \frac{1-\varepsilon_\infty}{\varepsilon_\infty A_\infty}\right) + \sum_{i=1}^{N}\left(2\frac{1-\varepsilon_{s,i}}{\varepsilon_{s,i} A_{s,i}} + \frac{1}{A_{s,i}}\right)}, \quad (\mathrm{F.1})$$

in which $\varepsilon, \varepsilon_{s,i}$, and ε_∞ are the emissivities of the sample, the ith radiation shield, and the environment, respectively, and likewise for the areas A of each of these components.

For tightly concentric shields placed close to the sample, where all of the gaps between the adjacent shields are small compared to the nominal diameter of the sample, we may obtain a linearized radiative heat transfer coefficient

$$h_{\text{rad'n-shields}} = \frac{4\varepsilon\sigma T_{\text{avg}}^3}{(1 + \varepsilon/\varepsilon_\infty - \varepsilon) + N(2\varepsilon/\varepsilon_s - \varepsilon)}, \tag{F.2}$$

where the numerator is Eq. (3.6) and the denominator represents a reduction factor. As an example, for a black sample and environment at room temperature, even a single shiny shield with $\varepsilon_s = 0.1$ will reduce the heat transfer coefficient from 6.1 W/m$^2 \cdot$K to 0.31 W/m$^2 \cdot$K, a reduction factor of 20.

For clarity, we focus on a one-shield scenario (Fig. F.1) to illustrate a key point in common experiments where the heated sample is much smaller than the environment. In this case, we require the diameter of the shield to be as close as possible to the diameter of the sample (Fig. F.1(a)), instead of the environment (Fig. F.1(b)). This is very important for the shield to actually reduce the radiation heat losses, which can be seen from Eq. (F.1) because the terms involving the shield(s) tend to vanish as $A_{s,i} \gg A$.

Note that the discussion above corresponds to a "thermally floating shield", which is thermally decoupled from the surroundings

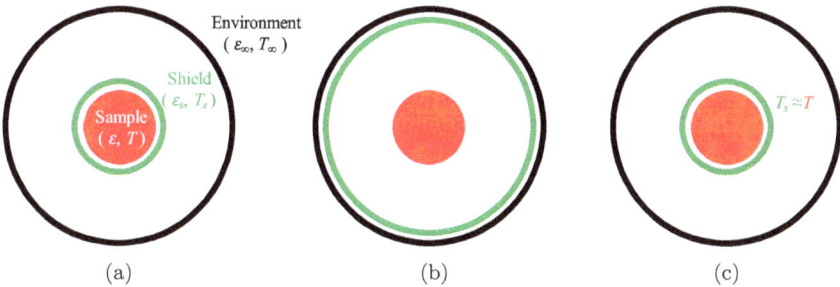

Fig. F.1 "Floating" (a, b) vs. "guarded" shields (c). (a) A preferred design, in which the diameter of the shield is chosen to be as small as possible, thereby minimizing the radiative heat losses from the heated sample. (b) A poor design, because a large shield is much less effective at reducing the radiative heat losses. The diameter effects in (a, b) are apparent from Eq. (F.1). (c) A "guarded" shield is actively controlled to maintain $T_s = T$. In principle this offers perfect thermal radiative insulation of the sample, because from fundamental thermodynamics once $T_s = T$, no heat can flow between the sample and the shield.

except for the obvious radiation interactions already contained in Eqs. (F.1) and (F.2). On the other hand, in both commercial cryostats and home-built customizations, the radiation shield(s) may be deliberately thermally anchored to some intermediate temperature, rather than floating. In the best case the shield temperature can even be tied (passively or even actively) to match the sample temperature, a so-called "guarded" or "driven shield" (Fig. F.1(c)) [18]. In principle this can completely eliminate the radiation losses because once $T_s = T$ no radiation heat can leave the sample, regardless of $h_{\mathrm{rad'n\text{-}shields}}$.

G. Material properties

When working on the thermal design stage of a new experiment, it is helpful to have order-of-magnitude estimates for the thermal properties expected of the sample, as well as collecting information about the thermal properties of the other relevant materials in the test structure. For example, in the heat spreader method as shown in Fig. 1.8 and discussed in Secs. 1.8, 2.3.2(F), and 3.2, making reasonable estimates of the thermal conductivities of the graphene flake and the bottom oxide layer, before the test structure is ever fabricated, is helpful for sizing the gaps between the thermometer lines.

In this section, we first briefly comment on the range and the underlying physics of three thermal properties of general interest, namely, the heat capacity, the thermal conductivity, and the thermal boundary resistance. For readers seeking more detail, we offer a few references commonly used in the nanoscale heat transfer community. We close by briefly commenting on the so-called "classical size effect", which can dramatically reduce the thermal and electrical conductivities of nanostructures as compared to their bulk counterparts.

G.1. *Heat capacity*

Around room temperature and above, the heat capacity for most fully dense materials is $\sim 3 \times 10^6$ J/m$^3\cdot$K, with variations of around a factor of 3 (see [19, Fig. 4.11]). This very helpful rule of thumb is relevant for all types of dense materials, including metals, dielectrics, and polymers, whether crystalline or amorphous, bulk or nanostructured. The underlying physics is the Dulong–Petit limit of the phonon heat capacity, which holds for temperatures higher than about half of the Debye temperature, and gives [20]

$$C = 3\eta k_B, \qquad (G.1)$$

where k_B is the Boltzmann constant and η is the number density of atoms. The key is that $\eta \sim 1 - 10 \times 10^{28}$ m^{-3} for the vast majority of fully dense solids.

For lower cryogenic temperatures, the Debye model is standard for the heat capacity. Note that even in metals, the total heat capacity is dominated by phonons down to $T \sim 1$ K, even though a metal's thermal conductivity is dominated by its electrons [15, 16, 20].

One helpful reference for heat capacity data of bulk materials is a classic handbook series, *Thermophysical Properties of Matter*, edited by Y. S. Touloukian [21]. In particular, Volumes 4–6 summarize numerous experimental results for metals and nonmetals, in solid, liquid, and gas phase. Other helpful references include the *Thermophysical Properties of Matter Database* (TPMD) [22] which builds on [21] and the *Landolt–Bornstein Database* [23].

G.2. *Thermal conductivity*

The thermal conductivity of common fully dense solids at room temperature spans only a fairly modest range as compared to the tremendous range of electrical conductivity. For example, a typical polymer may have k as low as $\sim$0.1 W/m·K, while the classic ultra-high-k material, diamond, has k around 2000 W/m·K. Thus the dynamic range of k is around four orders of magnitude. In comparison, the range of electrical conductivity in common materials spans over 25 orders of magnitude (see for example [19, Fig. 4.10]).

Pushing these extremes of thermal conductivity is one of the research drivers for the nanoscale heat transfer community [24]. At the high-k end, the challenge of measuring k of an individual carbon nanotube inspired the development of one of the signature measurement techniques, the suspended microfabricated device method [25–28], as discussed in Secs. 1.4 and 2.3.2(A). At the low-k end, a groundbreaking experiment demonstrated k in one direction of a fully dense solid to be ~0.06 W/m · K at room temperature, within a factor of two of the thermal conductivity of air [29]. With these developments in the nanoscale heat transfer community, the range of thermal conductivity at room temperature is now approaching five orders of magnitude.

References [21] (Volumes 1–3), [22] and [23] are again highly recommended resources for bulk thermal conductivity data. In addition, David Cahill has shared some of his group's experimental results for bulk materials and thin films online [30].

G.3.   *Thermal boundary resistance*

The thermal boundary resistance R_c'' is defined as the temperature difference across an interface per unit heat flux, and is the most common quantity used to characterize interfacial heat transfer. For high-quality, atomically intimate interfaces with at least one side nonmetallic, R_c'' is limited by the phonon contribution, which ranges from ~10^{-9} m^2 · K/W [31] to ~10^{-7} m^2 · K/W [32, 33]. High-quality metal–metal interfaces can have even lower R_c'', from ~10^{-9} m^2 · K/W to ~10^{-10} m^2 · K/W [34]. The effect of this boundary resistance is equivalent to the thermal conduction resistance of an amorphous silicon dioxide layer with thickness varying from several to hundreds of nanometers [35].

If the interface quality is less than outstanding, for example due to contamination, damage, oxidation, or other impurity phases, the transmissivity of energy carriers can be greatly degraded, and further increases in R_c'' by an order of magnitude or more are not surprising. Interfaces formed by joining two initially distinct pieces of matter almost always have orders of magnitude higher R_c'' [36], though exceptions are known [37].

A recent review of phonon-dominated boundary resistances for 34 materials pairs with high-quality interfaces suggests the following rule of thumb [33, 38] for the best-case (lowest) R_c'',

$$R_{c,\min}'' \approx \frac{1}{\alpha G_{\max}}, \qquad (G.2)$$

where

$$G_{\max} \approx \frac{1}{4}[C_{\mathrm{DLP}}v_s]_{\min}. \qquad (G.3)$$

Here α is an averaged transmission coefficient, empirically found to range from ~0.2 to 0.5 for high-quality interfaces [33, 38], C can be reasonably approximated with the Dulong–Petit heat capacity from Eq. (G.1), v_s is an averaged sound velocity, and Eq. (G.3) uses the minimum value after evaluating each material separately.

Readers interested in further understanding of the thermal boundary resistance at atomically intimate interfaces are referred to dedicated review articles [33, 39, 40] and additional discussion in [35, 41].

G.4. *Very brief note about size effects on the thermal and electrical properties of nanostructures*

As their characteristic lengths shrink to micro- and nanoscales, the thermal and electrical properties of nanostructures may differ greatly from those of their bulk counterparts, due to the truncation of the bulk mean free paths of the corresponding energy/charge carriers. This is the same fundamental mechanism as introduced in the vacuum chamber example of Fig. E.2, which generalizes to heat conduction by phonons, photons, and electrons. Understanding such size effects in nanoscale heat transfer is a major topic of contemporary research which is covered in depth in various specialist textbooks [15, 16, 42, 43] and related collections [41, 44–46]. Here we only give the briefest taste, to emphasize that the material properties used to design and interpret a micro-nanoscale experiment may differ tremendously from the handbook values.

As an example from the thermal domain, many of the techniques presented in this book rely on microfabricated silicon for some component of the measurement platform, so it is important to understand that k of silicon micro- and nanostructures can be much lower than bulk handbook values. For example, the in-plane k of Si films with thicknesses below $\sim$20 nm is over 5 $\times$ lower than k of an intrinsic silicon wafer [47]. Similarly, dramatic reductions are seen for Si nanowires (e.g., 10 $\times$ reduction for diameters below $\sim$30 nm [48]) and nanocrystalline Si (e.g., 5 $\times$ reduction in k for grain sizes below $\sim$100 nm [49]). Thermal conductivity reduction factors over a broader range of length scales may be found in [50, Fig. 5] and [51, Figs. 7 and 8]. These reduction factors are for k around room temperature, increasing by orders of magnitude at cryogenic T.

Similar issues arise in the electrical properties of microfabricated metals. For example, due to the additional scattering of electrons by the film surfaces and internal grain boundaries, the room temperature resistivity of a $\sim$200-nm wide, 50-nm thick gold heater line (see Fig. 3.4) is $\sim$3 $\times$ higher than handbook values [52]. This additional scattering also acts to reduce the temperature coefficient of resistance, causing this same line to have an α which is $\sim$2 $\times$ lower than handbook values. Both effects can be calibrated for any given microfabricated heater line, and indeed generally must be in order to succeed at the high accuracy resistance thermometry which underlies all of the techniques in this book.

H.   The lognormal distribution [53]

As discussed in the Monte Carlo method of uncertainty analysis (Sec. 4.4), for a parameter with large uncertainty it often makes more sense to assume it follows a lognormal rather than normal probability distribution, thereby ensuring its value can never be negative. The lognormal distribution for a random variable x is defined as

$$p(x) = \frac{1}{x\sigma\sqrt{2\pi}} \exp\left(-\frac{(\ln(x) - \mu)^2}{2\sigma^2}\right), \tag{H.1}$$

where $x > 0$. This distribution is specified by the two parameters μ and σ, which are the logarithmic mean and logarithmic standard deviation, respectively. It is important to note that the logarithmic mean is *not* the same as the conventional (arithmetic) mean, nor are the logarithmic and conventional standard deviations the same. However, they are easily related.

To see this, we first evaluate the (arithmetic) mean of the above distribution,

$$\bar{x} = \int xp dx = \exp\left(\mu + \frac{1}{2}\sigma^2\right). \tag{H.2}$$

Similarly, evaluating the standard deviation of Eq. (H.1),

$$SD = \sqrt{\int (x - \bar{x})^2 p dx} = \bar{x}\sqrt{e^{\sigma^2} - 1}. \tag{H.3}$$

It is more helpful to consider the relative standard deviation, which is purely a function of σ,

$$\frac{SD}{\bar{x}} = \sqrt{e^{\sigma^2} - 1}. \tag{H.4}$$

It is also insightful to consider the limiting behavior of the lognormal distribution for tight distributions. Considering $\sigma \ll 1$, to leading order we find

$$\bar{x} \approx e^{\mu} \tag{H.5}$$

and

$$\frac{SD}{\bar{x}} = \sigma. \tag{H.6}$$

These approximate forms are helpful for developing intuition about the relationships between logarithmic and conventional means and standard deviations.

References

[1] Manual of lock-in amplifier (Stanford Research SR850 DSP).
[2] R. Serway, *Principles of Physics*, 2nd ed. Saunders College, 1998.
[3] M. C. Wingert, Z. C. Y. Chen, E. Dechaumphai, J. Moon, J. Kim, J. Xiang and R. Chen, "Thermal conductivity of Ge and Ge–Si core–shell nanowires in the phonon confinement regime," *Nano Lett.* **11**(12), pp. 5507–5513, 2011.
[4] S. Sadat, E. Meyhofer and P. Reddy, "High resolution resistive thermometry for micro/nanoscale measurements," *Rev. Sci. Instrum.* **83**, p. 084902, 2012.

[5] P. M. Mayer, D. Lüerßen, R. J. Ram and J. A. Hudgings, "Theoretical and experimental investigation of the thermal resolution and dynamic range of CCD-based thermoreflectance imaging," *J. Opt. Soc. Amer. A* **24**(4), pp. 1156–1163, 2007.

[6] D. G. Cahill, "Thermal conductivity measurement from 30 to 750 K: The 3-omega method," *Rev. Sci. Instrum.* **61**(2), pp. 802–808, 1990.

[7] C. Dames and G. Chen, "1ω, 2ω, and 3ω methods for measurements of thermal properties," *Rev. Sci. Instrum.* **76**(12), p. 124902, 2005.

[8] D. R. White *et al.*, "The status of Johnson noise thermometry," *Metrologia* **33**, pp. 325–335, 1996.

[9] J. Crossno, X. Liu, T. A. Ohki, P. Kim and K. C. Fong, "Development of high frequency and wide bandwidth Johnson noise thermometry," *Appl. Phys. Lett.* **106**, p. 23121, 2015.

[10] S. Lee and D. G. Cahill, "Heat transport in thin dielectric films," *J. Appl. Phys.* **81**(6), pp. 2590–2595, 1997.

[11] J. H. Lienhard IV and J. H. Lienhard V, *A Heat Transfer Textbook*, 4th ed. Cambridge, MA: Phlogiston Press, 2012.

[12] R. E. Bentley, "The theory and practice of thermoelectric thermometry," in *Handbook of Temperature Measurement*. Springer, 1998.

[13] L. Lu, W. Yi and D. L. Zhang, "3ω method for specific heat and thermal conductivity measurements," *Rev. Sci. Instrum.* **72**(7), pp. 2996–3003, 2001.

[14] J. Kimling, S. Martens and K. Nielsch, "Thermal conductivity measurements using 1ω and 3ω methods revisited for voltage-driven setups," *Rev. Sci. Instrum.* **82**(7), p. 074903, 2011.

[15] C. Dames, "Microscale conduction," in *Heat Conduction*, 3rd ed. lead author Latif Jiji, Springer, 2009.

[16] G. Chen, *Nanoscale Energy Transport and Conversion*. New York: Oxford University Press, 2005.

[17] R. Siegel and J. R. Howell, *Thermal Radiation Heat Transfer*, 3rd ed. Washington, DC: Taylor & Francis, 1992.

[18] D. Salmon, "Thermal conductivity of insulations using guarded hot plates, including recent developments and sources of reference materials," *Meas. Sci. Technol.* **12**(12), pp. R89–R98, 2001.

[19] M. F. Ashby, *Materials Selection in Mechanical Design*, 3rd ed. Butterworth-Heinemann, 2005.

[20] C. Kittel, *Introduction of Solid State Physics*, 8th ed. John Wiley & Sons, 2005.

[21] Y. S. Touloukian, *Thermophysical Properties of Matter*. New York: IFI/Plenum.

[22] "Thermophysical properties of matter database (TPMD)," https://cindas data.com/products/tpmd [Online].

[23] "The Landolt–Bornstein database," http://materials.springer.com [Online].

[24] D. G. Cahill, "Extremes of heat conduction — Pushing the boundaries of the thermal conductivity of materials," *MRS Bull.* **37**(9), pp. 855–863, 2012.

[25] P. Kim, L. Shi, A. Majumdar and P. McEuen, "Thermal transport measurements of individual multiwalled nanotubes," *Phys. Rev. Lett.* **87**(21), p. 215502, 2001.

[26] L. Shi, D. Y. Li, C. H. Yu, W. Y. Jang, D. Kim, Z. Yao, P. Kim and A. Majumdar, "Measuring thermal and thermoelectric properties of one-dimensional nanostructures using a microfabricated device," *J. Heat Transfer* **125**(5), pp. 881–888, 2003.

[27] L. Shi, "Mesoscopic thermophysical measurements of microstructures and carbon nanotubes," PhD dissertation, University of California at Berkeley, 2001.

[28] D. Li, "Thermal transport in individual nanowires and nanotubes," PhD dissertation, University of California at Berkeley, 2002.

[29] C. Chiritescu, D. G. Cahill, N. Nguyen, D. Johnson, A. Bodapati, P. Keblinski and P. Zschack, "Ultralow thermal conductivity in disordered, layered WSe2 crystals," *Science* **315**(5810), pp. 351–353, 2007.

[30] D. G. Cahill, http://users.mrl.illinois.edu/cahill/tcdata/tcdata.html [Online].

[31] R. Costescu, M. Wall and D. Cahill, "Thermal conductance of epitaxial interfaces," *Phys. Rev. B* **67**(5), p. 54302, 2003.

[32] H.-K. Lyeo and D. Cahill, "Thermal conductance of interfaces between highly dissimilar materials," *Phys. Rev. B* **73**(14), p. 144301, 2006.

[33] C. Monachon, L. Weber and C. Dames, "Thermal boundary conductance: A materials science perspective," *Annu. Rev. Mater. Res.* **46**(1), pp. 433–463, 2016.

[34] R. B. Wilson and D. G. Cahill, "Experimental validation of the interfacial form of the Wiedemann–Franz law," *Phys. Rev. Lett.* **108**(25), p. 255901, 2012.

[35] D. G. Cahill, W. K. Ford, K. E. Goodson, G. D. Mahan, A. Majumdar, H. J. Maris, R. Merlin and S. R. Phillpot, "Nanoscale thermal transport," *J. Appl. Phys.* **93**(2), p. 793, 2003.

[36] F. P. Incropera, D. P. DeWitt, T. L. Bergman and A. S. Lavine, *Fundamentals of Heat and Mass Transfer*, 6th ed. John Wiley & Sons, 2007.

[37] J. Yang, Y. Yang, S. W. Waltermire, X. Wu, H. Zhang, T. Gutu, Y. Jiang, Y. Chen, A. A. Zinn, R. Prasher, T. T. Xu and D. Li, "Enhanced and switchable nanoscale thermal conduction due to van der Waals interfaces," *Nat. Nanotechnol.* **7**(2), pp. 91–95, 2012.

[38] R. B. Wilson, B. A. Apgar, W. P. Hsieh, L. W. Martin and D. G. Cahill, "Thermal conductance of strongly bonded metal-oxide interfaces," *Phys. Rev. B* **91**(11), pp. 1–7, 2015.

[39] E. Swartz and R. Pohl, "Thermal boundary resistance," *Rev. Mod. Phys.* **61**(3), pp. 605–668, 1989.

[40] P. E. Hopkins, "Thermal transport across solid interfaces with nanoscale imperfections: Effects of roughness, disorder, dislocations, and bonding on thermal boundary conductance," *ISRN Mech. Eng.* **2013**, Art. ID 682586, 2013.

[41] D. G. Cahill, P. V. Braun, G. Chen, D. R. Clarke, S. Fan, K. E. Goodson, P. Keblinski, W. P. King, G. D. Mahan, A. Majumdar, H. J. Maris, S. R. Phillpot, E. Pop and L. Shi, "Nanoscale thermal transport. II. 2003–2012," *Appl. Phys. Rev.* **1**(1), p. 11305, 2014.

[42] M. Kaviany, *Heat Transfer Physics*, 1st ed. Cambridge University Press, 2008.

[43] Z. M. Zhang, *Nano/Microscale Heat Transfer*, 1st ed. McGraw-Hill, 2007.

[44] S. Volz (ed.), *Microscale and Nanoscale Heat Transfer.* Springer, 2007.

[45] G. Chen (ed.), *Annual Review of Heat Transfer.* Begell House, 2014.

[46] S. Volz (ed.), *Thermal Nanosystems and Nanomaterials.* Springer, 2009.

[47] W. Liu and M. Asheghi, "Phonon-boundary scattering in ultrathin single-crystal silicon layers," *Appl. Phys. Lett.* **84**(19), pp. 3819–3821, 2004.

[48] D. Li, Y. Wu, P. Kim, L. Shi, P. Yang and A. Majumdar, "Thermal conductivity of individual silicon nanowires," *Appl. Phys. Lett.* **83**(14), pp. 2934–2936, 2003.

[49] Z. Wang, J. E. Alaniz, W. Jang, J. E. Garay and C. Dames, "Thermal conductivity of nanocrystalline silicon: Importance of grain size and frequency-dependent mean free paths," *Nano Lett.* **11**(6), pp. 2206–2213, 2011.

[50] E. S. Toberer, L. L. Baranowski and C. Dames, "Advances in thermal conductivity," *Annu. Rev. Mater. Res.* **42**(1), pp. 179–209, 2012.

[51] F. Yang and C. Dames, "Mean free path spectra as a tool to understand thermal conductivity in bulk and nanostructures," *Phys. Rev. B* **87**(3), p. 35437, 2013.

[52] W. Jang, Z. Chen, W. Bao, C. N. Lau and C. Dames, "Thickness-dependent thermal conductivity of encased graphene and ultrathin graphite," *Nano Lett.* **10**(10), pp. 3909–3913, 2010.

[53] W. Navidi, *Statistics for Engineers and Scientists*, 1st ed. Boston: McGraw-Hill, 2006.

Index

www.ingramcontent.com/pod-product-compliance
Lightning Source LLC
Chambersburg PA
CBHW050630190326
41458CB00008B/2212